Integrated Protected Area Management

Integrated Protected Area Management

Everett Parker

STATES
ACADEMIC PRESS
www.statesacademicpress.com

Published by States Academic Press,
109 South 5th Street,
Brooklyn, NY 11249, USA
www.statesacademicpress.com

Integrated Protected Area Management
Everett Parker

International Standard Book Number: 978-1-63989-292-1 (Hardback)

Cataloging-in-Publication Data

Integrated protected area management / Everett Parker.
p. cm.
Includes bibliographical references and index.
ISBN 978-1-63989-292-1
1. Protected areas. 2. Biodiversity conservation. 3. Public lands.
4. Environmental protection. I. Parker, Everett.
QH75 .I58 2022
333.951 6--dc23

Table of Contents

Preface

This book has been written, keeping in view that students want more practical information. Thus, my aim has been to make it as comprehensive as possible for the readers. I would like to extend my thanks to my family and co-workers for their knowledge, support and encouragement all along.

Protected areas are the areas which are conserved due to their recognized natural, cultural and ecological value. There are several types of protected areas such as marine protected areas, transboundary protected areas, national parks, nature reserves, wildlife sanctuaries, private protected area, etc. Integrated protected area management is the management of these protected areas for sustainable use of these regions. Assessment of management, planning and designing of these reserved areas is the focus of this domain. Principle aspects crucial for this management are theory and practice of designation, identification of priorities and community-based conservation. This book is compiled in such a manner, that it will provide an in-depth knowledge about the theory and practice of integrated protected area management. It studies, analyzes and upholds the pillars of this field and its utmost significance in modern times. Those in search of information to further their knowledge will be greatly assisted by this book.

A brief description of the chapters is provided below for further understanding:

Chapter - What is Protected Area Management?

Protected area management refers to the management of the conservation areas that aim at biodiversity preservation, providing habitat and protection from hunting of threatened and endangered species. This is an introductory chapter which will briefly introduce about protected areas management.

Chapter - Classification of Protected Areas

Protected areas can be classified into various sub-divisions such as national parks, transboundary protected areas, biosphere reserves, wilderness areas, habitat management areas, etc. This chapter closely examines the different protected areas to provide an extensive understanding of the subject.

Chapter - Marine Protected Areas

Marine protected areas include the protected areas of seas, oceans, estuaries, etc. These protected areas restrict human activities for protecting natural resources and conserving marine biodiversity. Marine reserves, marine parks, shark and whale sanctuaries, etc. are such protected areas. This chapter has been carefully written to provide an easy understanding of marine protected areas.

Chapter - Wildlife Sanctuaries

Wildlife sanctuaries refer to the natural habitat that protects the wildlife species from hunting, predation and poaching. It includes bird reserves, animal sanctuaries, protected forests, etc. The topics elaborated in this chapter will help in gaining a better perspective of wildlife sanctuaries.

Chapter - Threats to Protected Areas

The most common threats to protected areas include human-wildlife conflict, poaching, skinning, hunting, air pollution and acid rain, etc. This chapter delves into these threats to protected areas for an in-depth understanding of the subject.

Everett Parker

1

What is Protected Area Management?

Protected area management refers to the management of the conservation areas that aim at biodiversity preservation, providing habitat and protection from hunting of threatened and endangered species. This is an introductory chapter which will briefly introduce about protected areas management.

The IUCN's World Commission on Protected Areas has established six protected area management categories: strict nature reserve, wilderness area, national park, natural monument or feature, habitat/species management area, protected landscape/seascape, and protected areas with sustainable use of natural resources. Protected areas in the same category are managed based on similar objectives. For example, national parks are designated and managed to: "(a) protect the ecological integrity of one or more ecosystems for present and future generations, (b) exclude exploitation or occupation inimical to the purposes of designation of the area, and (c) provide a foundation for spiritual, scientific, educational, recreational and visitor opportunities, all of which must be environmentally and culturally compatible."

In addition to the IUCN categories, there are four other international designations of protected areas: Biosphere Reserve; World Heritage Site; Ramsar wetland, and marine. Biosphere Reserves serve three objectives: (a) conservation of landscapes, ecosystems, species, and genetic variation; (b) sustainable economic and human development; and (c) support for research, monitoring, education, and information exchange. World Heritage Sites possess outstanding cultural, natural, or mixed (combination of cultural and natural) values that are protected for the benefit of all humanity through cooperative international arrangements. The Convention on Wetlands is an intergovernmental treaty whose member countries protect Wetlands of International Importance in their territories. The United Nations Law of the Sea Treaty provides a legal framework for allocating rights to territorial seas and protecting marine-protected areas.

In 2011, there were 157,897 protected areas in the world covering 24,236,479 km^2, which amount to approximately 16% of the world's land surface area. Locations of protected areas cover the gamut from densely populated urban areas to unpopulated wilderness areas. Protected areas range in size from a few square kilometers to thousands of square kilometers. Most protected areas are publicly owned and managed. The level of protection afforded protected areas depends on many factors, including adequate funding, diligent law enforcement, effective management practices, and the strength of citizen support.

Protected areas are established and managed for a wide range of social, economic, and ecological values, including preserving biodiversity, particularly for threatened and endangered species; contributing to regional conservation strategies; maintaining diversity of landscape or habitat and of associated species and ecosystems; ensuring the integrity and long-term maintenance of specified conservation targets; providing a range of ecosystem goods and services that support sustainable use of natural resources; providing research and educational opportunities; maintaining and protecting cultural sites and traditional values; providing opportunities for tourism and recreation; carbon storage; disaster mitigation, soil stabilization; water supply for human and agricultural uses; and wilderness preservation.

Nature and Types of Protected Areas

The establishment of public protected areas was motivated by setting aside novel or awe-inspiring resources for public enjoyment. Biodiversity or ecosystem integrity was not a relevant concern, and the size of many protected areas was the smallest necessary to protect unique features. With the passage of time, these areas became increasingly valuable for protecting species and accomplishing other societal goals that required relatively unaltered landscapes. Today protected areas are being considered in national strategic plans, such as carbon sequestration for mitigating global warming or the creation of continental conservation networks, purposes for which they were not created. Outside the United States, the concept of protected areas has expanded to include traditional uses deemed worthy of preservation. For that reason, historic agricultural practices are maintained in certain protected areas to continue the interaction between people and the landscape and increase public support for those areas. Some countries have established parks with the explicit aim of attracting tourists and increasing national income and employment (e.g., African wildlife parks). This demonstrates the important contributions of protected areas to local and regional economies and that the motivation for establishing protected areas has changed.

Publicly owned protected areas are managed by federal, state, and local governmental agencies. For example, at the federal level in the United States, the National Park Service manages the National Park System, the Forest Service manages the National Forest System, several federal agencies manage units in the National Wild and Scenic Rivers System and National Wilderness Preservation System, the National Oceanic and Atmospheric Administration manages Marine Protected Areas, and the U.S. Fish and Wildlife Service manages the National Wildlife Refuge System. Each federal agency operates under the authority of an organic act that dictates how certain types of protected areas should be managed. States also designate and manage protected areas. For instance, the Missouri Natural

Areas System is an extensive statewide network of over 80 natural areas representing examples of the state's original natural landscape and outstanding biological and geological features. Local park commissions manage urban parks, which are typically oriented towards outdoor sports, picnicking, hiking, swimming, and enjoyment of nature. Nongovernmental organizations purchase land and conservation easements to protect natural values and fish and wildlife habitat. For example, The Nature Conservancy works with landowners, communities, cooperatives, and businesses to establish local groups that protect land through the use of land trusts, conservation easements, private reserves, and incentives. Nature Conservancy lands, which are located primarily in the United States, constitute the world's largest system of private nature reserves covering 60,703 km^2.

Threats to Protected Areas

As human populations, incomes, and leisure time increase, protected areas have come under greater pressure to serve more people and more diverse purposes often not envisioned when the protected areas were established. These human pressures vary greatly with country and the type of protected area but, in general, these dynamics has changed the role of protected area managers from caretakers to assertive advocates who manage their areas to provide benefits to local, regional, and national economies and minimize the impacts of threats to protected areas. Examples of such threats include rapid growth in visitation, changes in the mix of recreational and visitor uses (hunting, trapping, motorized/mechanized sports and transportation, hiking, camping, skiing, and air tours), development of roads and visitor facilities, spread of invasive species, residential and commercial development, legal resource extraction (logging, livestock grazing, farming, and energy development), illegal timber harvesting, wildlife poaching, air, water, and soil pollution, and climate change and variability. Some threats, such as increases in visitation and invasive species, are internal to protected areas, whereas others such as residential and commercial development and climate change, are external. In terms of the nature of the threats, increases in visitation and invasive species exert pressure on natural resources of protected areas and make their preservation more difficult. Landscape fragmentation, caused by residential and commercial development adjacent to protected areas, and climate change reduce the ecological integrity of protected areas, particularly their ability to serve as refugia for threatened and endangered species. External threats to protected areas are typically more difficult to resolve than internal threats because protected area managers have limited legal authority and are unwilling to control events external to their protected areas.

Management of Protected Areas

A major challenge facing protected area managers is how to manage those areas so as to achieve multiple management objectives. This challenge occurs for several reasons:

- First, certain management objectives (e.g., resource protection and visitor satisfaction) are somewhat incompatible.

- Second, there is considerable uncertainty about the natural and socioeconomic processes influencing protected areas, how internal and external threats influence those processes, and how different management actions are likely to influence the achievement of management objectives.
- Third, some management objectives, such as minimizing adverse impacts on visitors and the local economy, are supported by one group of stakeholders, whereas other objectives, such as natural and cultural resource protection, are supported by another group of stakeholders. Fourth, many protected areas face declining financial and human resources, which makes it more difficult to develop and implement effective management plans.

Two methods can be used by protected area managers to evaluate and make management decisions under uncertainty, namely, scenario planning and adaptive management (AM). Scenario planning requires protected area managers to develop and evaluate decisions for all combinations of unknown future events (e.g., future climate scenarios) and management actions. The scenario planning approach used by the U.S. National Park Service to manage national parks for climate change requires park managers to: 1) Choose the most probable climate scenario based on their subjective judgment; 2) Determine the most preferred management action for that climate scenario; and 3) Implement the most preferred management action.

A difficulty with scenario planning for climate change is that it requires protected area managers to choose the most probable climate scenario. This is problematic because scientists who develop climate scenarios have been unwilling or unable to assign probabilities to future climate scenarios. Thus, protected area managers need to develop management options for multiple scenarios.

AM postulates that "if human understanding of nature is imperfect, then human interactions with nature (e.g., management actions) should be experimental." Kohm and Franklin state that: "AM is the only logical approach under the circumstances of uncertainty and the continued accumulation of knowledge." AM acknowledges that protected area managers cannot accurately predict the outcomes of management actions because of environmental, scientific, organizational, community, and political uncertainties.

AM can be either passive or active. Application of passive AM to protected areas requires protected area managers to: 1) Formulate a predictive model of how a protected area responds to alternative management actions; 2) Select the most preferred management actions based on model predictions; 3) Implement and monitor those management actions; 4) Use monitoring results to revise the model; and 5) Adjust management actions based on the revised model. Advantages of passive AM are that it is relatively simple to use and usually less expensive to implement than active AM. A disadvantage of passive AM is that it does not produce statistically reliable information about the impacts of management actions on protected areas because it does not utilize experimental methods.

Application of active AM to protected areas requires protected area managers to:

- Formulate hypotheses about how various management actions influence a particular feature of protected areas (e.g., biodiversity) when there is uncertainty about how future events (e.g., climate change) interact with management actions to influence that feature.
- Establish experiments to test hypotheses about the efficacy of management actions over time.
- Use test results to determine the best management action to implement in an upcoming management period.
- Continue experiments and use experimental results to periodically retest the hypotheses to determine whether the best management action changes over time.

Active AM typically involves major investments in research, monitoring, and modeling; has prerequisites that are not always satisfied; and is more complicated to implement than passive AM.

Protected areas play increasingly important roles in our global society, including biodiversity conservation, providing ecosystem services, and supporting regional economies through recreational opportunities and tourism. These values of protected areas are reflected in the extensive distribution and diversity of protected areas and their increasing numbers. With a burgeoning global population and its demands on natural resources, protected areas will become even more critical to informed conservation planning and as scientific resources. Protected area managers must respond with vision and flexibility using new approaches such as scenario planning and AM.

Multiple uses: Sustainable Development in Protected Areas

Currently, almost no one argues that the overarching goal of PAs is conserving biodiversity and that any other management objective should be subject to biodiversity maintenance, improvement or restoration. There is, however, considerable debate on the weight of additional social and economic objectives in PA management, following the desirable yet vague concept of sustainable development.

Global agreements are giving PAs a starring role in many aspects other than biodiversity conservation. One extended mission included in its most updated definition is providing while preserving ecosystem goods and services and associated cultural features other than biodiversity. The most relevant goods and services provided by PAs include: raw materials, food, genetic, medicinal and ornamental resources, water purification, air quality regulation, erosion prevention, mitigation of extreme events, pollination, biological control, carbon sequestration, soil formation, primary production, and nutrient cycling. Lately, PAs are also increasingly conceived as optimum testing fields for monitoring global change, as well as useful and cost-effective means for climate change adaptation and mitigation. As seen, the "environmental targets" of PAs are quite ambitious. Not all PAs are able to provide all those goods and services, although some of them, especially the biggest ones, might.

Nevertheless, following the dominant tri- (or four-) dimensional concept of sustainability, PAs are also entrusted a high number of other social and economic roles, multiplying management objectives virtually to the infinite. Some of those roles are: scientific research, recreation and tourism, inspiration for art and culture, spiritual and cognitive development, aesthetic enjoyment, preservation of traditional cultures and practices, improving social welfare, enhancing environmental education and awareness, promoting peace and security, facilitating people's participation and governance, and boosting economic development at multiple scales.

Whereas attaining all those targets would be ideal for any PA, the challenge for PAs to do so is overwhelming. These goals are so wide and ambitious that, even in the cases where they are not directly opposed, it looks doubtful that any piece of land in the world could comply harmoniously with all of them.

Acknowledging the role of man within nature and its importance for the conservation of many species, habitats and ecological processes, especially in Europe where cultural landscapes are paramount, multiple uses of PAs sometimes becomes a "dogma" for PA planners, managers and policy-makers, irrespective of very different conditions and situations. Most conservation policies try to force sustainable development to happen, often through the zoning of PAs or the placing of tourist infrastructures, regardless of the fact that some activities may conflict with conservational ends. Such conflicts are frequent in densely populated areas between conservation and recreation activities.

Cultural features and traditions are an important and often neglected asset of protected areas.

Cultural, recreational, educational and economic values should be pursued in PA planning and management, as long as the key aim of PAs (i.e., conserving biodiversity) is never left behind to satisfy other possibly conflicting objectives. If social and economic objectives are compatible with biodiversity conservation or improvement, we should try to integrate those within planning and management. If they are not, their practice attainment should be considered no more than theoretical goodwill, no matter how temping development promises may look like. If poorly conceived, executed and regulated, even the best socioeconomic proposal might turn out to be the deadliest ecological error.

Therefore, predicted future trends on the prevailing designation of "multiple use" PAs (IUCN categories V and VI) in the years to come, although a good piece of news in itself, might entail subtle risks for effective conservation of resources.

The best solution to make most uses of PAs compatible would be designing a territorial model made of different types (regarding their priority function) of large, zoned PAs connected through biodiversity-oriented managed landscapes. However, limited resources and high land and visitor pressure, especially in densely populated areas, force us to put the preservation of ecosystem processes first and before any other use or consideration of PAs.

Role of Tourism in Protected Areas

Tourism is an extremely delicate issue regarding PA management. Whereas often seen and promoted as social and economic salvation for local communities, if unregulated it may, on the one hand, lead to the deterioration or destruction of the resources of the PA and, on the other, threaten local culture. If well planned and managed, however, tourism can provide a significant source of revenue for local populations and /or PA administrators, as well as increase visitor education and environmental awareness.

Visitors may be an intense pressure in protected areas.

Nevertheless, management and conservation problems are intrinsic to tourist visitation of PAs. These include: increasing use of resources, including land-use changes for developing tourism infrastructures in or around PAs; soil trampling and compacting; vegetation removal; animal disturbance; soil erosion; littering; noise making; introduction of alien species and varieties; damage to geological features, cultural sites, vegetation or public use infrastructure; increased fire risk; air pollution from transportation; animal road killing; changes in wildlife behavioural patterns due to human habituation, among others impacts. Some of these impacts can be so serious that they can compromise the long-term conservation of the resources being protected. Visitor impacts may also result in a reduced quality of recreational experiences and in conflicts among visitors.

Additionally, public use management competes for scarce funding with other management and conservation objectives of PAs, as it is a highly resource-demanding activity.

While zoning PAs and building tourist infrastructure might be useful at reducing impacts from tourism in PAs by directing main fluxes towards the least fragile zones of the PA, they cannot usually prevent invasion of more vulnerable zones. In fact, tourism infrastructure can attract even more visitors to the PA, and may result in its overcrowding and collapse.

As a result, although public use in PAs is a desirable and broadly-accepted concept, it should be strictly regulated or even prohibited in the cases when it is incompatible with conservation targets which must prevail in PAs.

Financial and Equity Aspects of Entrance Fees to Protected Areas

Impacts of visitation to PAs are numerous and serious. Whereas there is free access to most PAs in industrialised countries, no matter what land ownership is, entrance fees are common practise in other parts of the world.

There are a number of reasons in favour of charging a fee for accessing PAs. First and most important, entrance fees are effective visitor filters. On the one hand, they enhance the quality of visitors by discouraging the least interested visitors often having the poorest environmental awareness from entering the PA, thus helping prevent most unsocial behaviour. On the other, they potentially reduce the quantity of visitors, thus diminishing the human pressure on species, soil, vegetation, ecosystems, cultural features and infrastructure.

PAs provide multiple social goods and services related to improved health and well-being: Recreation, tourism, sport, relax, inspiration, cultural and aesthetic enjoyment, and spiritual and cognitive development. The fact that most of these goods and services have not market price does not mean their economic value is negligible; rather on the contrary. Thus, charging an amount of money to enjoy all these benefits from PAs does not seem exaggerate.

Additionally, paying an amount of money makes the visitor conscious of the value of the visited place. It gives him also a temporal "membership" to a "selected" group of people (club) with the right to respectfully enjoy the wonders of the PA.

Finally, most PAs are underfunded and budgetary restrictions will tend to be the rule more than the exception in the future due to financial constraints in the public sector which currently makes up most PA funding globally. Thus, provided that incomes from visitors are properly invested in PAs (in the same PA they are expended or in other PAs to support underfinanced or undervisited PAs), PA administrators could count on an additional source of financing to enhance management and conservation and to provide new employment opportunities.

Fee collection should be carefully planned and implemented. It must consider adequate equity prior to its implementation so none will be deprived of his or her right to enjoy PAs. Thus, discounts should be applied to low-income visitors upon appropriate certification. Dual-pricing policies charging foreign visitors higher than local ones are an equitable and socially acceptable option in developing countries.

Well-thought decisions should also be taken about visitors being charged repeatedly during their visits (for instance, for car parking, access to the PA, and /or visitation of public use infrastructure), and about the appropriate quantities to be charged in view of recovery costs, so no additional charges are imposed to the PA administration.

Parking fee.

Effective Management vs. Effective Conservation

There is no consensus on the effectiveness of PAs as a global strategy for biodiversity conservation. There are examples of conservation success and failure, although the latest are rarely reported. Lack of or inappropriate monitoring and assessment activities are common management deficiencies making the assessment of PAs often difficult, arbitrary or meaningless.

Therefore, whereas the protection of spaces for biodiversity conservation remains a valid and useful strategy to mitigate current environmental crisis, a general statement in favour of the actual effectiveness of PAs as a tool for biodiversity conservation is simplistic. Rather, PA effectiveness should be analysed case by case.

It is generally accepted that legal designation plus effective management result in the effective conservation of PAs. It is not always the case, however. There are important pressures and threats to the conservation of PAs and their resources that operate at a scale broader than PAs and that may, therefore, spoil brilliant management efforts, as they are outside the scope, means and competence of PA managers.

Impacts of global change are posing additional pressures on the conservation of biodiversity worldwide. Regional impacts can also cause severe effects on faraway zones regardless of their degree of protection or management effectiveness. Acid rain on land and oil spills on sea are good examples. As a result, a PA can be legally designated, well-planned and efficiently managed and see its resources being degraded, more so in the marine environment where connectivity is higher than on land.

PAs exchange matter, energy and information with their surroundings. These exchanges are vital to ecological processes underpinning biodiversity and other ecosystem goods and services. Shifts in nature, extent, direction and intensity of these interactions may also, however, result in significant alteration of the ecosystem structure and function, even if optimum PA management is in place. Therefore, PAs cannot be managed in isolation from the surroundings influencing those. As a result, nearly as much care should be taken when managing zones surrounding PAs as when managing PAs themselves.

Effective management can play an important role in PA conservation. It can also contribute to adaptation and mitigation of some regional or global pressures and threats, such as climate change, through PAs. However, the role of management in attaining long-term conservation of resources should not be overemphasized. PA managers can very rarely achieve effective conservation of managed sites on their own. In contrast, the importance of achieving sustainable development at all scales, integrating PAs in a wise and wider territorial planning cannot be stressed enough.

Assessing Protected Areas: Scales and Perspectives

For too many years, it was assumed that the resources harboured by PAs were effectively conserved just because they were legally protected. Regrettably, empirical and scientific evidences refute the previous assumption. Luckily, that vision is currently outdated and much literature has been written on "paper parks", although they still prevail in certain contexts.

Currently, the question is: what should be assessed? Or: how to measure PA effectiveness? Multivariate, socio-ecological systems such as PAs are difficult, costly and time-consuming to characterize, monitor and assess. Lack of basic knowledge increases the uncertainty of what to monitor, how and why?

As a result, developing a complete, ecologically-sound and meaningful-for-management assessment system has proved extremely complicated, despite several efforts developed worldwide. To start untangling this issue, a distinction between different concepts which are often used indiscriminately should be made. One is "management effectiveness" and the other is "PA effectiveness". Identifying both concepts would be the same as saying that all a PA can do to protect and conserve its resources is due to management. Once more, evidence does not seem to support this statement.

There are unprotected places well conserved due to lack of human visitation, accessibility and exploitation. There are paper parks enjoying a good conservation status for similar reasons. And there are legally-designated, well-managed PAs being degraded due to unsustainable socioeconomic contexts and regional or global pressures and threats.

Thus, PA effectiveness is not the same as management effectiveness. Stating so would put all the responsibility for the conservation status of PAs on managers with limited means, capacity and competences to cope with many factors outside their control. Thus, on the one hand, this identification is not precise and, on the other, it may neither be fair to PA managers.

Additionally, according to the "Management Effectiveness Evaluation Framework", protected area management effectiveness evaluation (PAME) should be target-driven. Therefore, if considered precisely, PAME could not be applied to most PAs worldwide lacking specific, clear, mensurable management objectives.

Therefore, "management effectiveness", which is an important part of "PA effectiveness", should only deal with factors directly linked to management which managers can control and address with sufficient knowledge, capacity and resources, while leaving out context variables they cannot manage or even influence. There are different conceptions of "PA effectiveness": From its identification with "management effectiveness", to "ecological effectiveness", "ecological integrity", or "sustainability".

Language precision is fundamental in science, but so far none has come up with an ideal, agreed term to designate overall "PA effectiveness" despite the wide literature on the topic: Are we measuring effectiveness?, performance?, functionality?, conservation potential?, sustainability?, success?, ecological integrity?, accomplishment? This uncertainty is because the core object being assessed is complex and integrates several different parameters (often environmental and also social and economic) not easily defined in one known-term.

Another question arises that further highlights the complexity of assessing PAs in an integrated and scientific manner: is "effectiveness" a common, "objective" concept? Is it site specific and thus, "subjective"? Or can it be both? The answer is not straightforward. There are different perspectives on PA effectiveness assessment closely linked to the scales considered and the objectives pursued with the assessment.

Scales are important when defining PA effectiveness. For example, "management effectiveness", "ecological effectiveness" and "ecological integrity" are usually conceived as "site-specific" due to the different ecological conditions, contexts and, therefore, management needs and objectives of every PA. In contrast, "sustainability assessment" proposes a common assessment and valuation method for PAs belonging to the same context (administrative, bio-geographical, or socioeconomic). Based on the most updated definition of PA, it assesses the likelihood that a PA can conserve its natural and associated cultural resources and ecosystem services in the long-term on relevant common parameters for management and decision-making. It is also a site assessment, but it allows comparison and prioritization among the individual PAs assessed as assessment is made upon the same parameters.

At a broader scale, the effectiveness of conservation networks should also be assessed using landscape metrics and indicators. Table below summarizes the different complementary approaches to assessing PA effectiveness.

Table: Complementary approaches to assessing protected area effectiveness.

Assessment	Scale	Object	Main target group
Ecological effectiveness	Individual PA (Site-specific)	Ecological parameters: ecosystems, species and the physical-chemical environment.	Scientists; PA managers.
Ecological Integrity	Individual PA (Site-specific)	Threats; Focal species, communities and ecological systems.	PA managers; decision-makers.
Management effectiveness	Individual PA (Site-specific; Comparable)	Management goals, objectives and strategies regarding: context, planning, inputs, processes, outputs and outcomes.	PA managers; PA network managers; decision-makers.
Sustainability / Effectiveness	Individual PA (Comparable)	State of conservation, planning, management, social and economic context, social perception and valuation, and threats to PAs.	PA network managers; PA managers; decision-makers.
PA network	Landscape; territory	Connectivity, biodiversity representativeness, gap analysis, etc.	PA network managers; decision-makers; territorial planners.

Protected Area Objectives vis à vis Management Objectives

Clear objectives are fundamental for adaptive management and, therefore, for effective PA conservation. Nevertheless, it is rare that clearly stated, science-based objectives be specified in any PA norm or management document and, when management objectives are set up, they are often vague, inadequate or even contradictory.

Thus, the establishment of clearly-defined objectives at different levels in the long, medium and short-terms remains one the most urgent needs for effective management and conservation of PAs.

PA objectives should be initially specified in the designation norm of each PA, according to its conservation and management characteristics. Although these objectives might change in the medium-term due to the evolution of natural systems, they are mostly stable throughout the time, as they portrait a long-term vision of the PA. They can be similar to or adapted from the general objectives of the different PA management categories proposed by the IUCN.

In contrast with these long-term "stable" objectives, medium-term specific objectives should be developed for each management planning period; i.e., the validity period of an average management plan of 4-10 years. These medium-term planning objectives should try to make the long-term objectives of the PA happen considering the different circumstances of the PA throughout the time.

Finally, management plans are often disaggregated into more operational annual workplans, whereby the medium-term objectives are made specific to the management means and to the reality of the PA through achievable and measurable short-term objectives. These short-term management objectives should also foresee (regarding means and probabilities of occurrence) and address sudden events not previously planned against which rapid action is required. Figure shows the conceptualization of the different types of objectives in PAs and their main characteristics.

Sensible objective definition leads to adequate planning and, if adequate means are available, to proper management of the PA. However, clear definition and precise forms of measurement of the different objectives are needed at all levels for management planning to be fully effective.

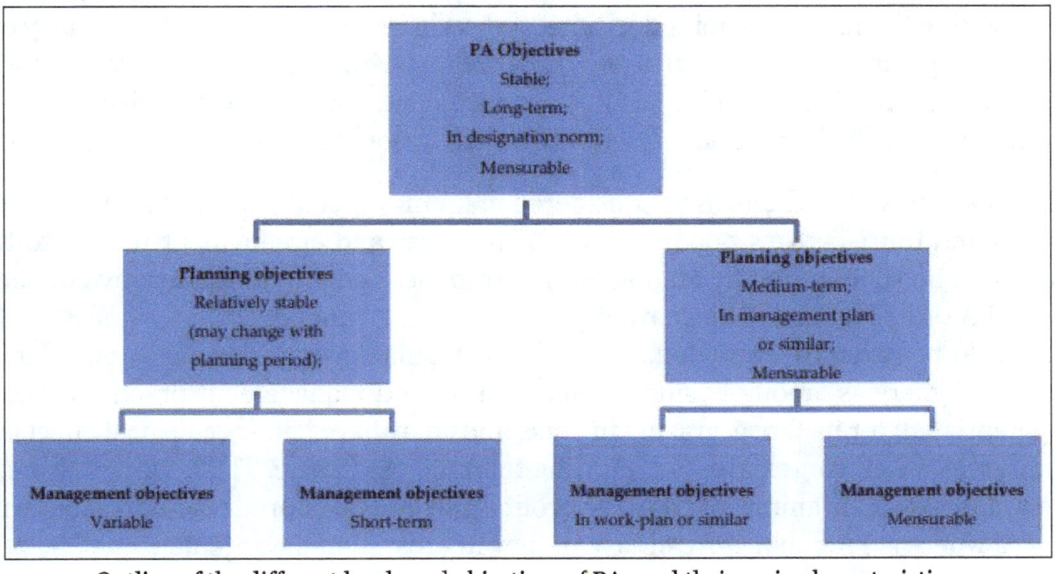

Outline of the different levels and objectives of PAs and their main characteristics.

Role of Society in Protected Area Management

PAs are portions of territory where multiple interests coexist and often collide. Although their main aim is to conserve biodiversity and the other ecosystem goods and services and associated cultural features, they include other environmental, social, economic and governance factors deserving attention. As such, they cannot and should not be managed in a unidirectional manner.

Neither decision-makers nor PA managers can cope with current challenges to PAs alone. An increasingly complex world requires increasingly innovative solutions, also regarding PA management. Thus, they should look for allies in other stakeholders if effective management and affordable enforcement is to be achieved.

Bottom-up, participatory approaches to PA management may be more resource and time consuming than technical top-down approaches, but they are generally accepted better by all stakeholders and, as a result, they are often more effective and enduring. Consensus results in decisions being more legitimate, and in fewer conflicts.

There are two priority groups whose close collaboration should actively be sought: scientists and local populations. Management measures can be more efficient, cost-effective and easily-implemented if they are based on sound science. Therefore, a stronger, deeper and more trustful relationship between PA managers and scientists is the first priority to enhance effective, participatory management. Local populations are also key to effective management. Their views and attitudes towards PAs and their management can determine the success of most management initiatives.

Building a positive attitude towards PAs by local populations consists not only in informing individuals or associations (NGOs, neighbours, etc.) on management issues, but also in consulting them, making joint decisions and seeking their input to actually carry out some management activities. Environmental volunteering can be a very adequate way of participation of local populations into management, and also an effective means of reinforcing the identity of residents with the values, resources and governance of the PA.

Other relevant stakeholders to be considered in PA management are businessmen, politicians and tourists. New opportunities for business and employment related to the PA should be sought, mainly among local enterprises, as long as no negative impacts on the PA or its resources are guaranteed. Therefore, economic activities affecting the PA should be carefully evaluated, planned and regulated in the management plan of each PA. Synergies among business, conservation and management practices should be potentiated; for instance, grazing in grasslands may have better ecological effects on biodiversity when complementing mowing than mowing alone. Thus, farmers can get free fodder for their animals and make profit by selling farm products associated with the PA, whereas PA managers can save money in ecosystem management.

Politicians should also be incorporated to valuating and managing PAs more actively. Their decisions can determine the future of a PA, including permission for aggressive activities within PAs or their surroundings, diminution or even disappearance of the protection status of a PA. Thus, they should be made aware on the importance of PAs and of their decisions for biodiversity and humans in the long-term. Social groups such as managers, scientists, local populations and NGOs can play an important role in lobbying for the cause of PAs.

Similarly, tourists should be made fully aware of their potential impacts on visited PAs through environmental education and volunteering, and public use infrastructure. Their collaboration with managers should be sought to obtain proper behaviour from them and, where possible, funding for the PA.

Keeping Pace with Science

PA knowledge and management do not run at the same speed. Communication and trust between scientists and managers can also be much improved. As a result, it is rare that sound, updated science be incorporated by managers and decision-makers and applied in everyday management. Therefore, common management practices are frequently based just on experience, habit, or at best, on partial scientific knowledge.

Sometimes, lack of scientific knowledge is due to overwork by managers. Too much field and administrative work leave them with little or no time or will to get updated on new management issues regularly. Lack of adequate funding, interest or training programmes can also erode appropriate scientific knowledge among managers. Thus, staff training is an overall need in many countries.

And yet widespread updated scientific knowledge is vital for management and conservation effectiveness. Resources devoted to conservation could be more successfully and wisely used if management practices were backed up by sound science. Staff training should be considered an integral part of the management cycle.

However, knowledge-transfer should not be unidirectional, from scientists to managers. The publication or communication of empirical management result and practices, be they successful or a failure, is also very important for advancing towards a more effective management of PAs and can provide science with interesting new research topics and experiences.

Presentations and workshops are useful training options.

Regular collaboration among scientists, managers and decision-makers is key for effective and efficient management. Current disconnection and suspicion should be overcome if more effective ways of conserving PA are to be sought. This collaboration can take the form of regular working-groups or workshops. Ideally, researchers should be incorporated to the regular PA managing staff be them restricted to specific PAs, covering a network of PAs, or both. Thus, actual management activities should be done considering both, sound science and best practice.

From Protected Areas to Protected Territories

Current in situ strategies for biodiversity conservation are not having enough success to stop or reverse biodiversity loss, and thus biodiversity and the other ecosystem goods and services continue to be degraded.

Most PAs, especially in Europe, are too small to maintain viable populations within their boundaries in the long-term. Under these circumstances, connectivity among PAs becomes fundamental.

Ecological networks are a conceptual and practical step forward with regard to the initial concept of PA as islands in the territory. However, ecological networks are not without theoretical and practical shortcomings. In addition, the current drivers causing environmental degradation are so intense than even ecological networks have been overcome by reality.

Enhanced connectivity of PAs through ecological corridors is thought to have increased PA conservation effectiveness. However, the wide range of scales taking part in ecological processes underpinning ecosystem goods and services make ecological corridors only partially effective, and only under some circumstances. Moreover, the zones surrounding PAs are becoming increasingly artificial as a result of human activities and are therefore not accomplishing properly their buffer target. As a result, the effective protection and wise management of whole landscapes/seascapes surrounding PAs can be as important as the protection and management of PAs themselves.

There is a growing consensus that effective biodiversity conservation cannot take place if complementary conservation scales broader than PA scale are not considered. Going from connected islands to connected landscapes is a desirable first step.

National conservation policies must be integrated and coordinated in order to be fully effective. The protection and management of isolated PAs, ecological networks or whole landscapes/seascapes alone is not sufficient for long-term conservation. Additional public conservation policies, mainly related to territorial planning, ecological restoration and environmental impact assessment (of projects, policies, plans and programmes), should complement the establishment and maintenance of PAs at the national level.

Full consideration of ecological processes should also be carefully considered and integrated in other public policies seriously affecting the steady provision of ecosystem goods and services such as housing, mining, fishing, transport or industry. Ecological considerations should always guide (and limit, when necessary) social and economic development policies and strategies if sustainable development is to be achieved. However, no matter how effective a national conservation policy is, its effectiveness will always be limited by regional or global trends, making local or national efforts only partially successful.

Future conservation strategies should be regional and global, as the driving forces pressing biodiversity are. Ecologically, it makes little sense planning and managing PAs restricted to local or national administrative boundaries. Thus, deeper and closer international collaboration is needed when planning for conservation.

Focus should change from PAs to protected territories of high conservation value following science-based criteria: Species richness, diversity, taxonomic uniqueness, degree of threat, endemism, unusual ecological or evolutionary phenomena, or notable provision of ecosystem goods and services. The ecoregional approach proposed by Olson et al. seems a promising and adequate scale to plan and manage conservation from an ecological perspective. This is the conservation approach followed by relevant conservation organizations, such as the Nature Conservancy.

Management should be much more collaborative and imply different stakeholders from different sovereign authorities. This form of closer international cooperation can have positive impacts in other public policies such as human development or defence.

Even if conservation and management strategies become global and effective, they will only be a limited solution to current and increasingly unsustainable human development trends. Thus, if an enduring solution to the environmental crisis (and thus, sustainable development) is to be achieved, global determined and courageous measures must be taken in order to counter current population, development and consumption trends. All public policies must take into account the bio-physical limits of the Biosphere and they should be planned and implemented accordingly, considering the long-term consequences of today's decisions.

PROTECTED AREAS MANAGEMENT EFFECTIVENESS

The evaluation of management effectiveness is generally achieved by the assessment of series of criteria (represented by carefully selected indicators) against agreed objectives or standards. Management effectiveness evaluation is defined as the assessment of how well protected areas are being managed – primarily the extent to which management is protecting values and achieving goals and objectives. The term management effectiveness reflects three main 'themes' in protected area management:

- Design issues relating to both individual sites and protected area systems.
- Adequacy and appropriateness of management systems and processes.
- Delivery of protected area objectives including conservation of values.

IUCN-WCPA has developed a management effectiveness evaluation framework which provides a consistent basis for designing evaluation systems for protected areas. Components of the IUCN framework include design of systems and individual protected areas (context and planning), appropriateness of management systems and processes (inputs and processes), and delivery of protected area objectives (outputs and outcomes). These components are divided into six elements, elaborated below, each comprising a number of evaluation indicators to assess management effectiveness.

Assessment needs to be made in the context of the protected area, so first assessments need to gather data on issues relating to the areas values, threats and opportunities, stakeholders, and the management and political context. Management starts with planning of strategies needed to fulfil the vision, goals and objectives of protection and to reduce threats. To put these plans in place and meet management objectives, managers need inputs (resources) of staff, money and equipment. Management activities are implemented according to accepted processes (i.e. best practices); which produce outputs by completing activities outlined in work plans. The end result of management is the achievement of outcomes, i.e. reaching the goals and objectives set for the biological conservation, economic development, social sustainability or cultural heritage of the protected area.

Design and Issues

1. Context: Assessment of Importance, Threats and Policy/Cultural Environment

The context review, although not an analysis of management, helps managers put their decisions in context and helps prioritise action based on biological, cultural and political information. The context assessment can help managers answer the following questions:

- Why is the protected area important?
- What are the threats facing the protected area?
- Is the government supportive of the protected area?
- What is the role and effect of stakeholders on protected area management?

2. Planning: Assessment of Protected Area Design and Planning

The assessment of management planning draws from the findings of the context assessment, Planning involves understanding the direction and objectives of management and deciding on the strategies that are required to achieve these, within the context of the protected area's status and characteristics. The planning assessment can therefore help managers to answer three questions:

- Is the legal status and tenure of the protected area is clear?
- How do the protected area's characteristics (e.g. size and shape) influence management?
- Is there an adequate management planning process?

Appropriateness of Management Systems and Processes

1. Inputs: Assessment of Resources needed to Carry out Protected Area Management

Allocating funds and staff time and developing infrastructure to fulfil management needs should be linked to and, in large part, directed by planning decisions. Although protected area management plans rarely provide specific commitments of funds and staff, they establish the basis for short-term or annual operational planning in which decisions about allocation of resources (budgets) are made. An assessment of resources allows managers to identify shortfalls in staff, funds and equipment in relation to planned activities. Assessment considers the resources required for effective management, measures these against what is available and looks at the effectiveness of resource use. The assessment thus has to address two important questions:

- Has the protected area got the resources needed to meet its management objectives?
- Are resources used in the best way?

2. Processes: Assessment of the Way in which Management is Conducted

The use of the best possible management practices is essential for effective protected area site or system management. Assessment can show if best practice is being applied in day-to-day management and if management practices can be or need to be improved. The process assessment asks:

- Are agreed policies and procedures for management in place?
- Are the best systems and standards of management being followed?
- How can the management practices be improved?

Delivery of Protected Area Objectives

1. Outputs: Assessment of the Implementation of Management Programmes, Actions and Services

The results of management activity can be considered in two ways – the outputs and outcomes. First, there are the direct outputs produced by management, which usually consist of a set of products or services (e.g. an area of controlled burn, number of invasive species eradicated, length of track maintained, numbers of guided walks conducted or numbers of anti-poaching patrols). The output assessment looks at what managers and their staff have been doing and whether the targets set in management plans or annual work programmes have been met. An output assessment therefore asks questions such as:

- Has the number or level of products and services been delivered?
- Have planned actions, tasks and strategies been implemented?

2. Outcomes: Assessment of the Extent to which Objectives have been Achieved

The outcome assessment considers impacts of management on the broad objectives set for the protected area. It looks beyond the implementation of immediate management activities (i.e. the outputs) to the longer term success of the protected area by attempting to answer the question:

- Has management resulted in the achievement of the objectives of, and desired outcomes for, the protected area?

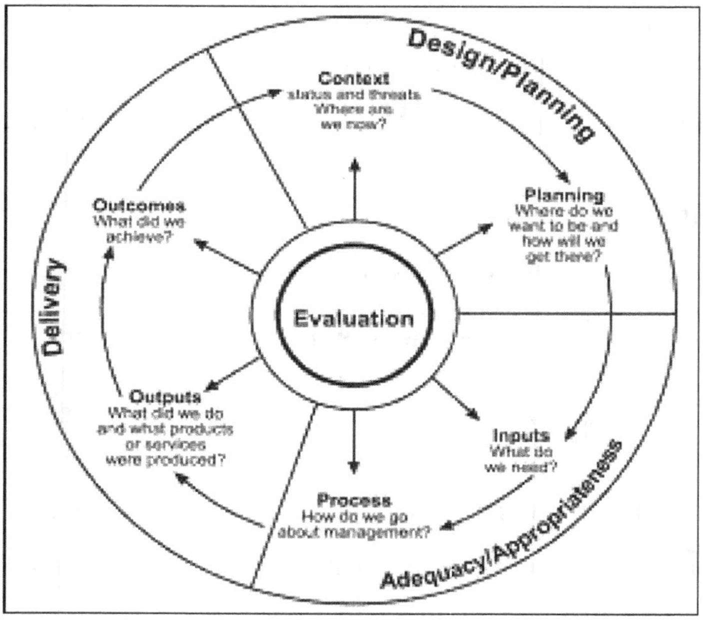

Approaches to outcome evaluation ideally involve long-term monitoring of the condition of the biological and cultural resources of the system/site, socio-economic aspects of use and the impacts of the management on local communities. In the final analysis, outcome evaluation is the true test of management effectiveness. Even if other aspects of management are highly effective, a protected area will clearly not be effective if it loses its core values.

Table: The IUCN-WCPA Framework for assessing management effectiveness of protected areas and protected area systems.

	Design		Appropriateness/Adequacy		Delivery	
Element	Context	Planning	Inputs	Process	Outputs	Outcomes
Evaluation Focus	Importance, Threats and policy/ cultural environment.	Design and Planning.	Adequacy of resource needed to manage.	How management is conducted.	Implementation of management programmes and actions.	Extent to which objectives have been achieved.
Criteria that are assessed	Values.	Protected area legislation and policy.	Resources available for management.	Sustainability of management process.	Results of management actions.	Effects of management in relation to objectives.
	Threats.	Protected area or systems.				

Since the development of WCPA framework in 2000, technical experience increased rapidly resulting in a range of assessment systems based upon the framework. There are now three basic approaches:

- In-depth, evidence based assessments aimed at building monitoring systems and long-term understanding of management in an individual protected area, such as the Enhancing our Heritage system being developed for World Heritage sites.
- System-wide peer–based assessment developed specifically for use on a system-wide scale such as the WWF RAPPAM system and the systems developed in Finland, Catalonia (Spain) and New South Wales (Australia).
- Scorecard expert–based assessments.

There are four major steps in assessing protected area management effectiveness: (i) Getting started, (ii) Gathering data, (iii) Analyzing results and (iv) Integrating into capacity assessments.

Getting Started

Important considerations for getting started include: (i) Who – setting up an organizing committee consisting of key agency staff, NGOs, donors and local people, who have detailed knowledge about policies, field level conditions and interagency dynamics as well as organizational, communication, facilitation and analytical skills, (ii) Where – the scope of assessment, either an entire system or an individual protected area. (iii) How - gathering information through workshops, direct interviews using questionnaires based on WCPA framework. For example WWF- RAPPM questionnaire contain more than 100 questions and relies on qualitative scoring by workshop participants.

Gathering Data

This includes four parameters of information collection:

(i) Significance and Vulnerability gathering:

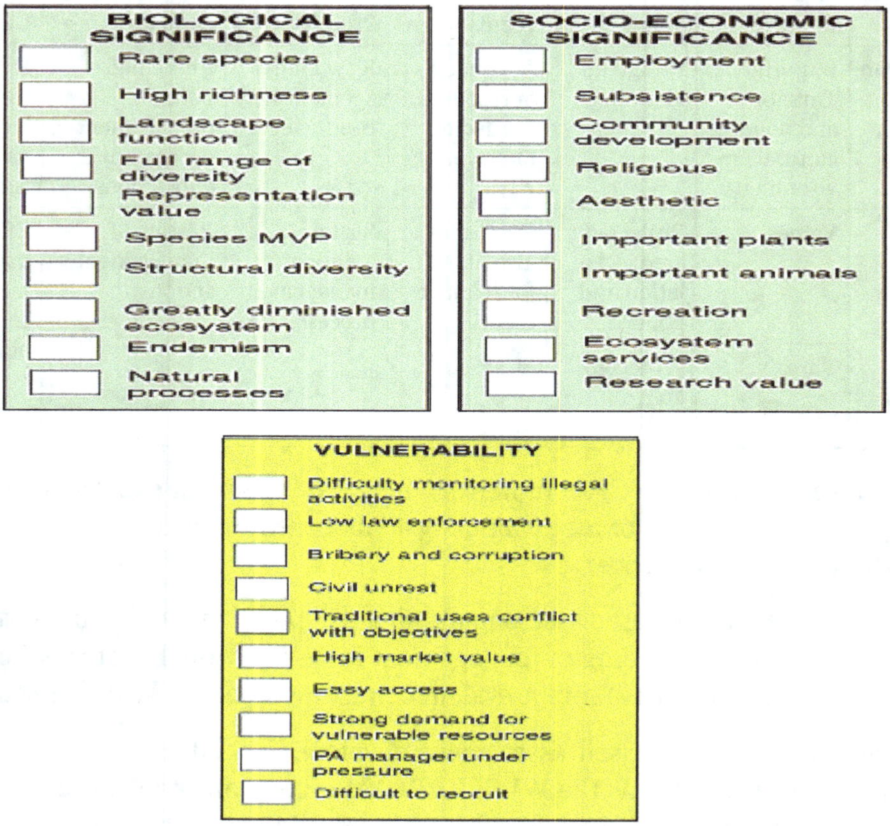

(ii) Threats and Pressures include: Any human activity that impairs biodiversity; Applies to existing and future threats; scored by extent, severity and permanence; and can incorporate multiple data sources.

(iii) Management Elements:

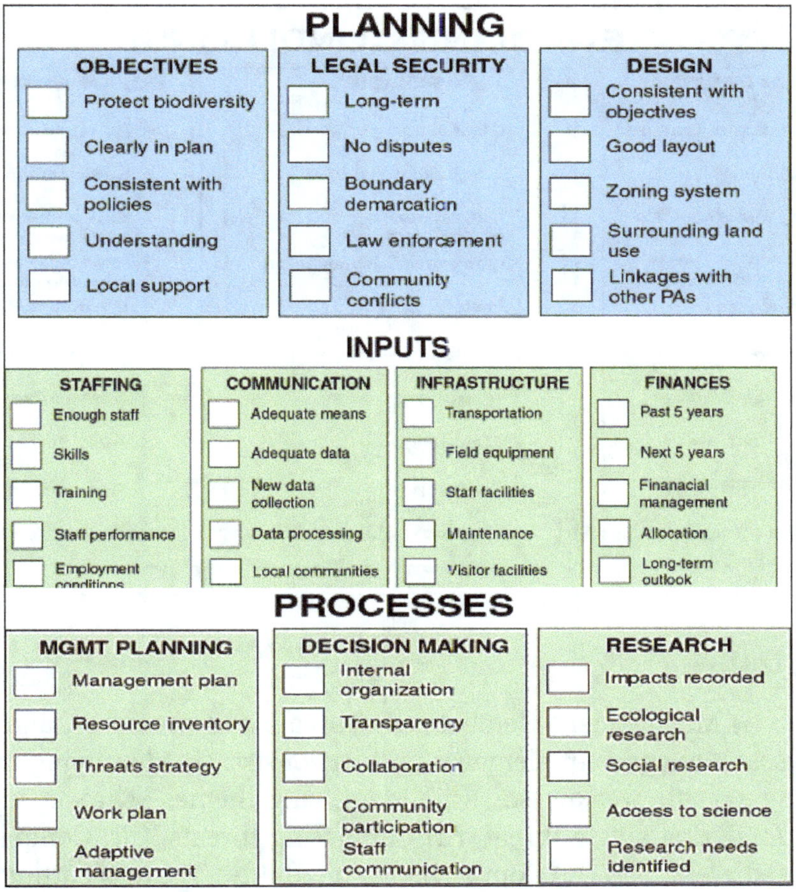

(iv) Assessment Methodologies:

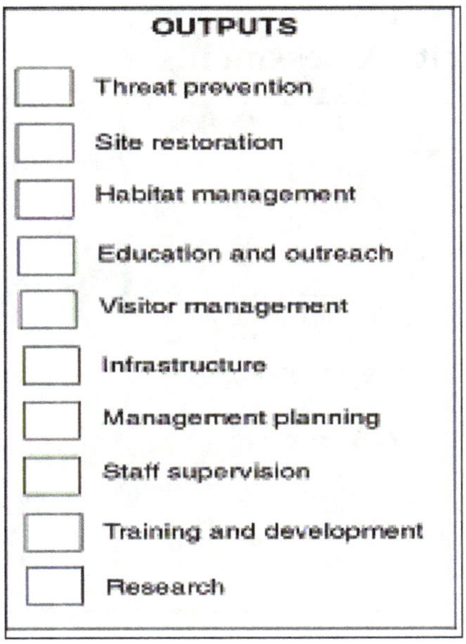

(v) System – Wide Enabling Environment:

ENABLING CONDITIONS

PA SYSTEM DESIGN	PA POLICIES	POLICY ENVIRONMENT
☐ Representation of systems	☐ Clear vision	☐ PA-related laws
☐ No extinctions	☐ Adequate coverage	☐ Sufficient funding
☐ Examplary intact systems	☐ Demonstrated commitment	☐ Environmental protection integrated into laws
☐ High conservation value sites conserved	☐ Comprehensive inventory	☐ Inter-departmental communication
☐ Natural processes	☐ Historic variability	☐ Environmental education
☐ Transition areas	☐ Restoration targets	☐ Law enforcement
☐ Successional diversity	☐ Ongoing research	☐ Sustainable land use
☐ Biodiversity rich areas conserved	☐ Gap analysis	☐ Array of mechanisms
☐ High endemism areas conserved	☐ Staff capacity	☐ Adequate training
☐ Layout and configuratoin	☐ Routine evaluation	☐ Civic dialogue

Analyzing Data

This includes: (i) Management elements (ii) Threats and pressures, and (iii) Cross–cutting analyses. Management elements in turn consist; (i) Identifying broad trends, (ii) Identifying specific weaknesses, (iii) Identifying themes. Threats and pressures includes: (i) Analyzing single threat, (ii) Comparing threats, (iii) Comparing threats across protected areas, and (iv) Comparing vulnerability. (i) Cross cutting analysis includes understanding threat and significance, and (ii) Understanding the relationship between effectiveness, threat and significance.

Integrating into Capacity Assessments

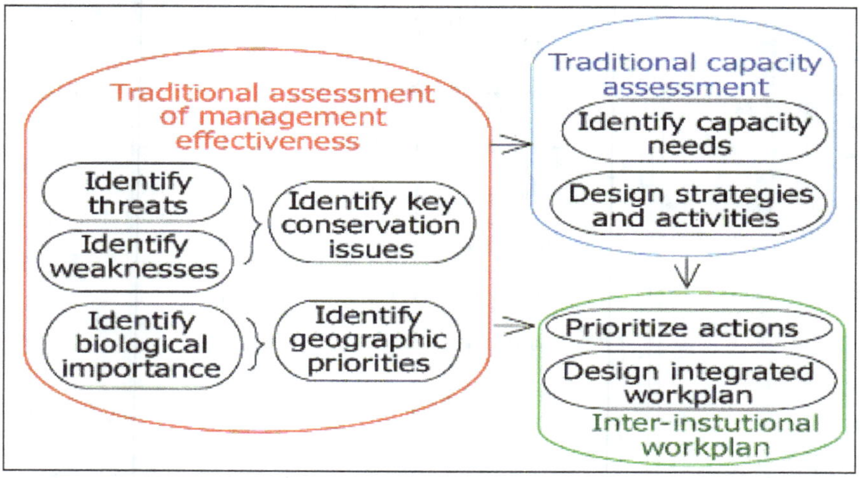

PROTECTED AREA MANAGEMENT PLANNING

A protected area management plan is the central mechanism to apply legislation and policy. A Management Plan will:

- Identify key issues.
- Define role and significance of an area within a system.
- Set out policy and zoning for protection, development and management of resources and attributes.
- Ensure that development and management are compatible with environmental protection.
- Provide a basis for ongoing monitoring of PA development.
- Facilitate communication and understanding within organization and outside.
- Provide continuity.

The creation, implementation and revision of protected area management plans both depends on and can facilitate meaningful stakeholder participation. It has been widely recognized that the parks and reserves that do not engage with local communities will become a focal point for conflict and discontent, and are at great risk of failing for this reason. Involving stakeholders in management planning acknowledges their desire and right to participate in decisions that affect them; facilitates understanding, enriches protected area management, improves and smooths decision making, promotes partnerships, stewardship and cooperation, and removes potential conflict.

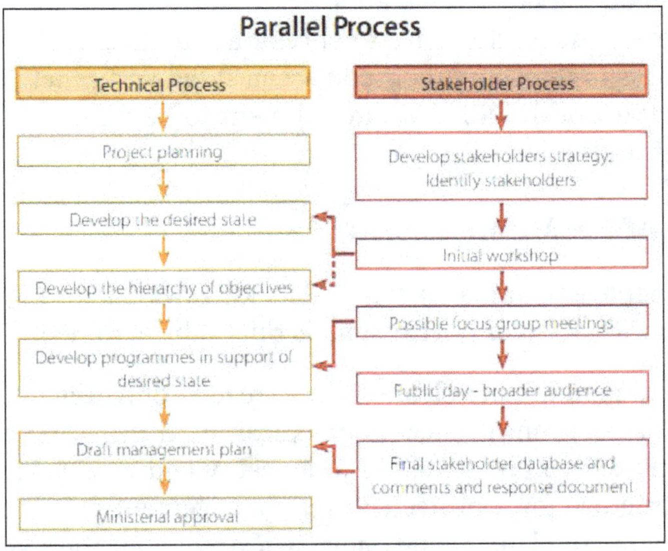

The parallel technical and stakeholder processes of protected area management planning.

The formulation of a protected area management plan does not end with the creation of the document itself – creating of a plan must be followed by plan implementation, but also by a circular process of monitoring and evaluation that will assess whether objectives remain appropriate and are being met by their associated action plans.

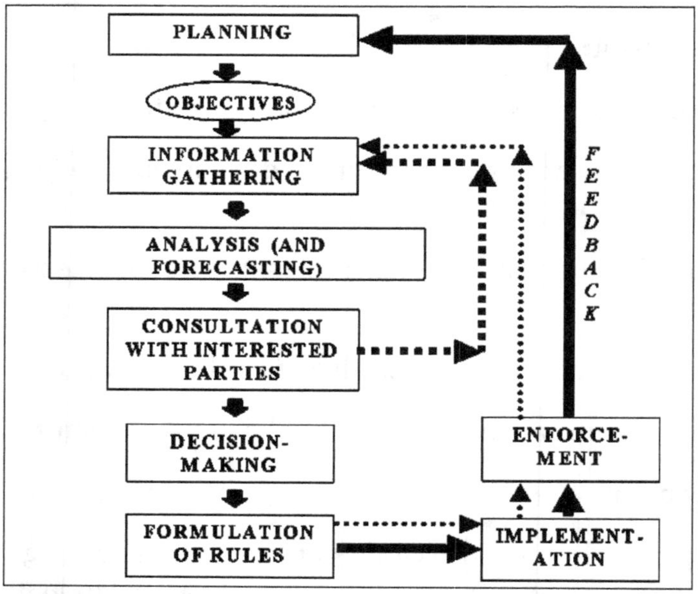

Flowchart of the Protected Area Management Planning process.
Items in bold are those specifically dealt with in workshop sessions.

The history of effective formal management plans for protected areas is relatively recent. For example, even though Yellowstone National Park, USA, the first National Park in the world, was established in 1872, it was only in 1916 that the US Department of Interior created the National Park Service to "promote and regulate areas such as reserves and National Parks". It took another 70 years before the US National Park Service in 1988 set out instructions for the preparation of general Management Plans for all National Parks in the USA. Their policy can stand as a general guideline for any protected area in the world: "No new development to be undertaken without an approved Management Plan".

The Structure of PA Management Plans

A protected area management plan is forward looking, with a 20-25 year time horizon, but sets objectives with a 5-year horizon and is thus subject to review every ~5 years.

A good management plan will be: Clear – easy to read and written in the language of its primary target audience; concise; accurate; logical; motivating; precise and practical; focused and effective. It will be seated in practicality and must be a living document. There is no set structure for a protected area management plan, each agency or organisation will formulate plans to suit their own needs, but there are some common elements: Context, Vision, Objectives, and Actions.

Context

The Vision provides a long–term perspective that is unlikely to change over the short to medium term. It will identify the main purposes of the protected area for the next 25 years and describes the ideal condition, state or appearance of the protected area that management is intended to achieve. A Vision needs to consider the Social, Technical, Environmental, Economic, and Political values of the protected area (i.e. V-STEEP). A Vision Statement should not be too long, and it should inspire.

Vision Statements Address the Question: "What should our Natural Area be Like?"

They may include components such as:

- An area that is rich in biological diversity and natural character.
- A protected and appropriately managed area for the flora and fauna it supports.
- A protected and appropriately managed area for future generations to enjoy.
- A healthy wetland/dune/farm/creek/bushland area which sustains a diversity of indigenous flora and fauna.
- A community that is aware and actively participating to restore and protect natural areas and habitats.

Objectives

Objectives follow on from Vision and are set in terms of the target future. They are specific statements of intentions that set out measurable outcomes, but not how to achieve these. Objectives should be SMART, i.e. Specific, Measurable, Achievable, Realistic, and Time-related, e.g. 5-10 years. The wording of an objective may not include specific targets, but it should be able to be translated into measurable outcomes. There may be specific objectives relating to different issues or functions within the protected area, including such topic headings as: Biodiversity and Heritage; Sustainable Tourism; Building Co-operation; Effective Park Management; Corporate Support and others or similar.

Zoning

Zonation within a protected area is both a mechanism for delimiting appropriate activities, and a means of defining and focussing the objectives guiding various management actions. There is a wide variety of zonation schemes possible and no simple all-purpose model. Each protected area will have different values to be protected and different activities or degrees of public access to be regulated.

One example of zonation categories is provided by the Planning and Management Policy, Ontario Provincial Parks, Canada:

- Natural Environmental Zone Natural, cultural and aesthetic landscapes; low-intensity recreation.
- Development Zone Main access and facilities and services for a wide range of day use/camping.
- Wilderness Zone Protection of significant natural and cultural features; minimum development.
- Natural Reserve Zone Significant Earth and Life Science features; minimum development.
- Historical Zone Significant historical features; minimum development for research and visitors.
- Access Zone Staging area where facilities support the use of NR and WI zones; less development than NE and HI zones.

Actions

Actions are the activities by which a given objective will be achieved. There can be be several actions for each objective. Actions tell a manager what must be done and how, so they must be measurable, clearly defined and achievable.

From the Actions it becomes possible to:

- Prioritise,
- Identify indicators to measure action,
- Set targets,
- Define responsibilities,
- Set the time frame,
- Determine the required budget.

Necessarily Actions are:

- Activity based.
- Linked to salary and wages = performance.
- And may relate to a number of sub-components:
    - Conservation development,

- Equipment,
- Staff,
- Biodiversity and heritage,
- Sustainable tourism,
- Building cooperation and constituency building,
- Effective park management,
- HR,
- Park Planning and development,
- Overheads.

WORLD WIDE FUND FOR NATURE

WWF in full stands for World Wide Fund for Nature and it is an international organization committed to conservation of the environment. In North America it is called the World Wildlife Fund.

Panda logo for the Switzerland-based
World Wildlife Fund (World Wide Fund for Nature).

In 1960 a group of British naturalists—most notably biologist Sir Julian Huxley, artist and conservationist Peter Scott, and ornithologists Guy Mountfort and Max Nicholson—led an effort to establish an organization that protected endangered species and their habitats. The following year the World Wildlife Fund was founded; the international name was subsequently changed to World Wide Fund for Nature in 1989, although in the United States and Canada it retained the founding name. The organization's distinctive panda logo was created by Scott. In the face of growing environmental threats over the ensuing years, the WWF's activities expanded in scope. Today its mission statement is threefold: To conserve the world's biological diversity, to ensure that the use of renewable natural resources is sustainable, and to promote the reduction of pollution and of wasteful consumption. The organization has long included both conservationists and businesspeople with the intention of combining solid scientific data with well-managed action. It also seeks cooperation between nongovernmental organizations, local governments, and local populations. The WWF works closely with the World Conservation Union and has formed partnerships with the United Nations, the World Bank, and the European Union.

The WWF provides money for conservation initiatives around the world. These include programs focused on individual species, forests, and freshwater and marine issues as well as climate change and responsible international trade. The group has also been involved in efforts to provide a safe and sustainable habitat for the world's peoples, both urban and rural, including clean water, clean air, healthful food, and rewarding recreation areas. Among the WWF's notable achievements is its use of debt-for-nature swaps, in which an organization buys some of a country's foreign debt at a discount, converts the money to local currency, and then uses it to finance conservation efforts. The WWF's first successful debt-for-nature swap took place in 1987 in Ecuador.

At the beginning of the 21st century, the WWF was active in more than 100 countries and had more than five million supporters. The organization's international headquarters are in Gland, Switz., and it has more than 90 offices around the world.

GREEN BELTS

Green Belts are a planning tool that helps maintain a ring of productive rural landscape around a city.

By designating an area 'green belt' it provides an additional layer of legal authority that allows local authority planners to reject requests for planning permission for most uses that would change the basic rural character of the area. In short, it stops urban areas from expanding by saying to people – you can buy and sell this land to your heart's content, but you can't expect to build housing estates or industrial estates on it; it stays as countryside.

The first green belt in Britain was demarcated around London. (Green belts aren't 'built' or signposted – but marked as lines on planners' maps – with an inner edge closer to an urban centre and the outer edge, beyond which more lenient planning regulations apply again).

The 1930's had seen rapid expansion of many towns and cities as, first, public transport (buses and trains) and later, increasing numbers of private cars enabled people to live further from their place of work and become 'commuters'. The result was rapid 'urban sprawl' with unregulated urban expansion. This was not seen as a desirable trend, so the 1947 Town and Country Planning Act introduced the first green belts.

Many other cities followed London's example and applied to government to establish green belts around their own urban centres. At present there are 14 green belts in England, 1 in Wales, 10 in Scotland and 30 in N. Ireland.

Purpose

The intentions behind green belts vary with each city and region of the country. Often more than one of the following applies:

- To prevent the merging of distinct urban centres by restricting urban sprawl.
- To preserve rural landscapes for leisure and recreation within reasonable reach of urban residents.
- To protect the distinctive character of historic and cultural cities.
- To conserve valuable high-quality agricultural land in food production.
- To encourage developers to re-use derelict brownfield urban land rather than greenfield agricultural land.

Criticisms

- They encourage 'leap-frog' development – pushing new developments to just beyond the outer edge of the green belt.
- They drive up house-prices within the urban area by restricting new house building.
- With a growing population it becomes harder to justify constraining urban growth.
- They can deter industrial expansion and commercial development.

Strengths

- Popular with planning departments – they can plan more effectively by designating (or not) land as green belt.

- They are easy to administer and manage.
- They are popular with rural communities covered by them as fields and views are less likely to be developed.
- They have helped stimulate the regeneration of much inner city land in recent decades.

Alternatives

There are variations to encircling green belts. Some cities have 'green buffers' to segregate two urban areas and prevent them merging, 'green wedges' bring greenery close to the heart of the urban area but permit alternate zones of urban growth, and others designate a 'green heart' with urban growth encouraged around them.

The sustainable cities movement emphasises the need for 'green spaces' within and beyond cities as necessary for a high quality of life. As far as Britain's green belts are concerned there are frequent calls to remove green belt designation over certain areas to allow necessary urban expansion to take place.

But others say this is 'foot in the door' land-use change that will make it more difficult to preserve the remaining green belt. Some argue from a political angle that green belts serve the interests of affluent suburban and rural-urban fringe residents, at the expense of the less affluent seeking affordable housing.

Others say that restricting the land for commerce and business in a globalised world will simply send companies abroad. Yet others argue that the regeneration of many central city areas on once-derelict industrial land couldn't have taken place without green belt policy forcing developers to use second choice land.

HABITAT MANAGEMENT

Once protected, areas must often be managed in order to maintain the threatened species within them. Management may involve the removal of alien species. It can also involve restoring natural ecological processes to the area. Original fire and flooding regimes are examples of such processes, and they are often controversial because human actions can alter them significantly.

Fire Control

Despite the often valid reasons for suppressing wildfires, the practice can change vegetation dramatically and sometimes harm species in the process. Human activities have changed fire regimes across large areas of the planet, including some biodiversity hot spots. Getting the fire regimes right can be essential for conserving species.

An example of a species for which the control of fire regimes has proven both possible and essential is Kirtland's warbler. This endangered species nests only in the Upper Peninsula of Michigan, an exceptional case of a bird species with a tiny geographic range well outside the tropics. The bird places its nest in grasses and shrubs below living branches of jack pines that are between 5 and 20 years old. The region's natural wildfires originally maintained a sufficient area of young jack pines. As elsewhere, modern practices suppressed fires, and the habitat declined. The birds are also susceptible to cowbirds, which are parasitic egg-layers. Active management with prescribed fires to ensure that there are always jack pines of the right age, together with the removal of cowbirds, has steadily increased the population of warblers since the early 1990s, when it had included only about 200 singing males. By 2015 there were about 2,360 singing males.

Flood Control

In much the same way that human actions suppress fire regimes, they also control water levels, and the resulting changes can have important consequences for endangered species. An example of a species so affected is the Cape Sable seaside sparrow found in the Florida Everglades. The Everglades once stretched from Lake Okeechobee in the north to Florida Bay in the south. Water flowed slowly over a wide area, and its levels varied seasonally: summer rains caused the levels to rise in late summer and early fall; then the dry season dropped the levels to their lowest in late May. Under this natural regime, some areas were continuously flooded for many years, drying out in only the driest years, whereas others were flooded for only a few months each year. It is in these drier prairies that the sparrows nest from about the middle of March until the water floods their nests in the summer.

Water-management actions have diverted the flow of water to the west of its natural path, making the western part of the Everglades unnaturally wet in some years during the bird's nesting season. Unnatural flooding during the four years that followed 1992 reduced the bird population in the west to less than 10 percent of its 1992 level. In the east the prairies have become unnaturally dry and so have become subject to an increased number of fires, which has jeopardized the sparrow population there. It is clear that rescuing the Cape Sable seaside sparrow requires the restoration of the natural water levels and flows within the Everglades and the consequent return to natural fire regimes.

Habitat Restoration

Once a habitat has been destroyed, the only remaining conservation tool is to restore it. The problems involved may be formidable, and they must include actions for dealing with what caused the destruction. Restorations are massive ecological experiments; as such, they are likely to meet with different degrees of success in different places. Restoration of the Everglades, for example, requires restoring the natural patterns of water flow to thousands of square kilometres of southern Florida.

A case history of habitat restoration comes from the Midwestern United States. In

Illinois, natural ecosystems cover less than 0.1 percent of the state, so restoration is almost the only conservation tool available. North Branch is a 20-km (12-mile) strip of land running northward from Chicago along the north branch of the Chicago River. Early in the 20th century, it was protected from building but later abandoned. Beginning in the 1970s, a group of volunteers first cleared out introduced European buckthorn, removed abandoned cars, and planted seeds that they had gathered from combing the tiny surviving remnants of original prairies in places such as along railways and in cemeteries. The initial effort to replant the prairie species failed, however; animals ate the small growing plants. The solution was to restore the natural fire regime, although controlled burns—setting fires safely near homes—posed a difficult technical challenge. But once it was accomplished, the results were immediate and dramatic. The original prairie plants flourished and the weeds retreated, although with an important exception. Under the native trees grew nonnative thistles and dandelions.

The original habitats locally called barrens constituted a visually striking and ecologically special habitat. Restoring them was a particular challenge, and the main conservation problem was finding the right mix of species. One recommendation was to use remnant barrens as models, but the North Branch volunteers rejected them as being too degraded. In the early 20th century, naturalists had speculated that barrens were special because they lacked some characteristic prairie species and, at the same time, had their own distinctive species. The volunteers examined habitat descriptions in old treatises of local plants looking for such species, many of which certainly would now be scarce. The key discovery was a list of barrens plants published by a country doctor in 1846. The scientific names had changed in 150 years, but, by tracking them through the local literature, the volunteers found many of their putative barrens-specific plants on the doctor's list. In possession of this vital information, the group succeeded in restoring barrens sites that by 1991 held 136 native plants, including whole patches consisting of species that, until restoration, had been locally rare.

The example above illustrates two rules of ecological restoration. The first is that one must be able to save all the component species of the original ecological community. The second, which is illustrated by the doctor's list, is that one needs to know which species belong and which do not. Without the right species mix, restorers must constantly weed and reseed.

References

- WWF, topic: britannica.com, Retrieved 02 June, 2020
- Green-belts, geography-reference: tutor2u.net, Retrieved 10 July, 2020
- RAPPAM Ervin, J. (2003). WWF: Rapid Assessment and Prioritization of Protected Area Management (RAPPAM) Methodology. WWF, Gland, Switzerland
- Habitat-management, science-conservation-ecology: britannica.com, Retrieved 13 August, 2020
- Guidelines for Applying Protected Area Management Categories; IUCN: Gland, Switzerland, and Cambridge, UK, 2008

2

Classification of Protected Areas

Protected areas can be classified into various sub-divisions such as national parks, transboundary protected areas, biosphere reserves, wilderness areas, habitat management areas, etc. This chapter closely examines the different protected areas to provide an extensive understanding of the subject.

STRICT NATURE RESERVE

Strict nature reserve are protected areas that are strictly set aside to protect biodiversity and also possibly geological/geomorphological features, where human visitation, use and impacts are strictly controlled and limited to ensure protection of the conservation values. Such protected areas can serve as indispensable reference areas for scientific research and monitoring.

Primary Objective

To conserve regionally, nationally or globally outstanding ecosystems, species (occurrences or aggregations) and geodiversity features: these attributes will have been formed mostly or entirely by non-human forces and will be degraded or destroyed when subjected to all but very light human impact.

Other Objectives

- To preserve ecosystems, species and geodiversity features in a state as undisturbed by recent human activity as possible.
- To secure examples of the natural environment for scientific studies, environmental monitoring and education, including baseline areas from which all avoidable access is excluded.
- To minimize disturbance through careful planning and implementation of research and other approved activities.
- To conserve cultural and spiritual values associated with nature.

Distinguishing Features

The area should generally:

- Have a largely complete set of expected native species in ecologically significant densities or be capable of returning them to such densities through natural processes or time-limited interventions.
- Have a full set of expected native ecosystems, largely intact with intact ecological processes, or processes capable of being restored with minimal management intervention.
- Be free of significant direct intervention by modern humans that would compromise the specified conservation objectives for the area, which usually implies limiting access by people and excluding settlement.
- Not require substantial and on-going intervention to achieve its conservation objectives.
- Be surrounded when feasible by land uses that contribute to the achievement of the area's specified conservation objectives.
- Be suitable as a baseline monitoring site for monitoring the relative impact of human activities.
- Be managed for relatively low visitation by humans.
- Be capable of being managed to ensure minimal disturbance (especially relevant to marine environments).

The area could be of religious or spiritual significance (such as a sacred natural site) so long as biodiversity conservation is identified as a primary objective. In this case the area might contain sites that could be visited by a limited number of people engaged in faith activities consistent with the area's management objectives.

Role in the Landscape/Seascape

Category Ia areas are a vital component in the toolbox of conservation. As the Earth becomes increasingly influenced by human activities, there are progressively fewer areas left where such activities are strictly limited. Without the protection accompanying the Ia designation, there would rapidly be no such areas left. As such, these areas contribute in a significant way to conservation through:

- Protecting some of the earth's richness that will not survive outside of such strictly protected settings.
- Providing reference points to allow baseline and long-term measurement and monitoring of the impact of human-induced change outside such areas (e.g., pollution).

- Providing areas where ecosystems can be studied in as pristine an environment as possible.
- Protecting additional ecosystem services.
- Protecting natural sites that are also of religious and cultural significance.

What makes Category Ia Unique?

Allocation of category is a matter of choice, depending on long-term management objectives, often with a number of alternative options that could be applied in any one site. The following box outlines some of the main reasons why Category Ia may be chosen in specific situations vis-à-vis other categories that pursue similar objectives.

Category Ia differs from the other categories in the following ways:

Category Ib	Category Ib protected areas will generally be larger and less strictly protected from human visitation than category Ia: although not usually subject to mass tourism they may be open to limited numbers of people prepared for self-reliant travel such as on foot or by boat, which is not always the case in Ia.
Category II	Category II protected areas usually combine ecosystem protection with recreation, subject to zoning, on a scale not suitable for category I.
Category III	Category III protected areas are generally centred on a particular natural feature, so that the primary focus of management is on maintaining this feature, whereas objectives of Ia are generally aimed at a whole ecosystem and ecosystem processes.
Category IV	Category IV protected areas protect fragments of ecosystems or habitats, which often require continual management intervention to maintain. Category Ia areas on the other hand should be largely self-sustaining and their objectives preclude such management activity or the rate of visitation common in category IV. Category IV protected areas are also often established to protect particular species or habitats rather than the specific ecological aims of category Ia.
Category V	Category V protected areas are generally cultural landscapes or seascapes that have been altered by humans over hundreds or even thousands of years and that rely on continuing intervention to maintain their qualities including biodiversity. Many category V protected areas contain permanent human settlements. All the above are incompatible with category Ia.
Category VI	Category VI protected areas contain natural areas where biodiversity conservation is linked with sustainable use of natural resources, which is incompatible with category Ia. However large category VI protected areas may contain category Ia areas within their boundaries as part of management zoning.

Issues for Consideration

- There are few areas of the terrestrial and marine worlds which do not bear the hallmarks of earlier human action, though in many cases the original human inhabitants are no longer present. In many cases, category Ia areas will therefore require a process of restoration. This restoration should be through natural processes or time-limited interventions: if continual intervention is required the area would be more suitable in some other category, such as IV or V.

- There are few areas not under some kind of legal or at least traditional ownership, so that finding places that exclude human activity is often problematic.
- Some human actions have a regional and global reach that is not restricted by protected area boundaries. This is most apparent with climate and air pollution, and new and emerging diseases. In an increasingly modified ecology, it may become increasingly difficult to maintain pristine areas through non-intervention.
- Many sacred natural sites are managed in ways that are analogous to 1a protected areas for spiritual and cultural reasons, and may be located within both category V and VI protected areas.

WILDERNESS AREA

A wilderness area is a region where the land is in a natural state; where impacts from human activities are minimal—that is, as a wilderness. It might also be called a wild or natural area. Especially in wealthier, industrialized nations, it has a specific legal meaning as well: As land where development is prohibited by law. Many nations have designated wilderness areas, including Australia, Canada, New Zealand, South Africa and the United States.

The World Conservation Union (IUCN) classifies wilderness at two levels, Ia (Strict Nature Preserves) and Ib (Wilderness areas).

Most scientists and conservationists agree that no place on earth is completely untouched by humanity, either due to past occupation by indigenous people, or through global processes such as climate change. Activities on the margins of specific wilderness areas, such as fire suppression and the interruption of animal migration, also affect the interior of wildernesses.

There are twelve wilderness areas in the Sami native region in northern Finnish Lapland. They are intended both to preserve the wilderness character of the areas and further the traditional livelihood of the Sami people. This means e.g. that reindeer husbandry, hunting and taking wood for use in the household is permitted. As population is very sparse, this is generally no big threat to the nature. Large scale reindeer husbandry has influence on the ecosystem, but no change is introduced by the act on wilderness areas. The World Commission on Protected Areas (WCPA) classifies the areas as "VI Protected area with sustainable use of natural resources".

Since 1861, The French Waters and Forests Military Agency (Administration des Eaux et Forêts) put a strong protection on what was called the "artistic reserve" in Fontainebleau State Forest. With a total of 1 097 hectares, it is known to be the first World nature reserve.

Then in the 1950s, Integral Biological Reserves (Réserves Biologiques Intégrales, RBI) are dedicated to interference-free ecosystem evolution, on the contrary of Managed Biological reserves (Réserves Biologiques Dirigées, RBD) where a specific management is applied to conserve vulnerable species or threatened habitats.

Integral Biological Reserves occurs in French State Forests or City Forests and are therefore managed nowadays by the National Forests Office. In such reserves, all harvests coupe are forbidden excepted exotic species elimination or track safety works to avoid fallen tree risk to visitors (already existing tracks in or on the edge of the reserve).

At the end of 2014, there were 60 Integral Biological Reserves in French State Forests for a total area of 111 082 hectares and 10 in City Forests for a total of 2 835 hectares.

There are seven wilderness areas in New Zealand as defined by the National Parks Act 1980 and the Conservation Act 1987 that fall well within the IUCN definition. Wilderness areas cannot have any human intervention and can only have indigenous species re-introduced into the area if it is compatible with conservation management strategies.

In the United States, a Wilderness Area is an area of federal land set aside by an act of Congress. Human activities in wilderness areas are restricted to scientific study and non-mechanized recreation; horses are permitted but motorized vehicles and equipment are not.

In Western Australia, a Wilderness Area is an area that has a wilderness quality rating of 12 or greater and meets a minimum size threshold of 8,000 hectares in temperate areas or 20,000 hectares in arid and tropical areas. A wilderness area is gazetted under section 62(1)(a) of the Conservation and Land Management Act 1984 by the Minister on any land that is vested in the Conservation Commission of Western Australia.

The IUCN has published their view on wilderness protection, governance and management, Wilderness Protected Areas: Management Guidelines for IUCN Category 1b Protected Areas.

NATIONAL PARK

A national park is a park in use for conservation purposes. Often it is a reserve of natural, semi-natural, or developed land that a sovereign state declares or owns. Although individual nations designate their own national parks differently, there is a common idea: The conservation of 'wild nature' for posterity and as a symbol of national pride.

An international organization, the International Union for Conservation of Nature (IUCN), and its World Commission on Protected Areas (WCPA), has defined "National Park" as its Category II type of protected areas. According to the IUCN, 6,555 national parks worldwide met its criteria in 2006. IUCN is still discussing the parameters of defining a national park.

Bogd Khan Uul National Park, Mongolia.

While this type of national park had been proposed previously, the United States established the first "public park or pleasuring-ground for the benefit and enjoyment of the people", Yellowstone National Park, in 1872. Although Yellowstone was not officially termed a "national park" in its establishing law, it was always termed such in practice and is widely held to be the first and oldest national park in the world. However, the Tobago Main Ridge Forest Reserve, and the area surrounding Bogd Khan Uul Mountain are seen as the oldest legally protected areas, predating Yellowstone by nearly a century.

National parks are almost always open to visitors.

Manuel Antonio National Park.

In 1969, the IUCN declared a national park to be a relatively large area with the following defining characteristics:

- One or several ecosystems not materially altered by human exploitation and occupation, where plant and animal species, geomorphological sites and habitats are of special scientific, educational, and recreational interest or which contain a natural landscape of great beauty.

- Highest competent authority of the country has taken steps to prevent or eliminate exploitation or occupation as soon as possible in the whole area and to effectively enforce the respect of ecological, geomorphological, or aesthetic features which have led to its establishment.
- Visitors are allowed to enter, under special conditions, for inspirational, educative, cultural, and recreative purposes.

Beech trees in Mallard Wood.

In 1971, these criteria were further expanded upon leading to more clear and defined benchmarks to evaluate a national park. These include:

- Minimum size of 1,000 hectares within zones in which protection of nature takes precedence.
- Statutory legal protection.
- Budget and staff sufficient to provide sufficient effective protection.
- Prohibition of exploitation of natural resources (including the development of dams) qualified by such activities as sport, hunting, fishing, the need for management, facilities, etc.

While the term national park is now defined by the IUCN, many protected areas in many countries are called national park even when they correspond to other categories of the IUCN Protected Area Management Definition, for example:

- Swiss National Park, Switzerland: IUCN Ia – Strict Nature Reserve.
- Everglades National Park, United States: IUCN Ib – Wilderness Area.
- Victoria Falls National Park, Zimbabwe: IUCN III – National Monument.
- Vitosha National Park, Bulgaria: IUCN IV – Habitat Management Area.
- New Forest National Park, United Kingdom: IUCN V – Protected Landscape.
- Etniko Ygrotopiko Parko Delta Evrou, Greece: IUCN VI – Managed Resource Protected Area.

While national parks are generally understood to be administered by national governments (hence the name), in Australia national parks are run by state governments and predate the Federation of Australia; similarly, national parks in the Netherlands are administered by the provinces. In Canada, there are both national parks operated by the federal government and provincial or territorial parks operated by the provincial and territorial governments, although nearly all are still national parks by the IUCN definition.

In many countries, including Indonesia, the Netherlands, and the United Kingdom, national parks do not adhere to the IUCN definition, while some areas which adhere to the IUCN definition are not designated as national parks.

In 1810, the English poet William Wordsworth described the Lake District as a "sort of national property, in which every man has a right and interest who has an eye to perceive and a heart to enjoy." The painter George Catlin, in his travels through the American West, wrote during the 1830s that the Native Americans in the United States might be preserved "(by some great protecting policy of government) in a magnificent park. A nation's Park, containing man and beast, in all the wild and freshness of their nature's beauty."

Starting in 1735 the Naples government undertook laws in order to protect Natural areas, which could be used as a game reserve by the royal family; Procida was the first protected site; the difference between the many previous royal hunting preserves and this one, which is considered to be closer to a Park rather than a hunting preserve, is that Neapolitan government already considered the division into the present-day wilderness areas and non-strict nature reserves.

First Efforts: Hot Springs, Arkansas and Yosemite Valley

Yosemite Valley, Yosemite National Park.

The first effort by the U.S. Federal government to set aside such protected lands was on 20 April 1832, when President Andrew Jackson signed legislation that the 22nd United States Congress had enacted to set aside four sections of land around what is now Hot Springs, Arkansas, to protect the natural, thermal springs and adjoining mountainsides for the future disposal of the U.S. government. It was known as Hot Springs

Reservation, but no legal authority was established. Federal control of the area was not clearly established until 1877.

John Muir is today referred to as the "Father of the National Parks" due to his work in Yosemite. He published two influential articles in The Century Magazine, which formed the base for the subsequent legislation.

President Abraham Lincoln signed an Act of Congress on 1 July 1864, ceding the Yosemite Valley and the Mariposa Grove of giant sequoias (later becoming Yosemite National Park) to the state of California. According to this bill, private ownership of the land in this area was no longer possible. The state of California was designated to manage the park for "public use, resort, and recreation". Leases were permitted for up to ten years and the proceeds were to be used for conservation and improvement. A public discussion followed this first legislation of its kind and there was a heated debate over whether the government had the right to create parks. The perceived mismanagement of Yosemite by the Californian state was the reason why Yellowstone, at its establishment six years later, was put under national control.

First National Park: Yellowstone

Grand Prismatic Spring in Yellowstone National Park.

In 1872, Yellowstone National Park was established as the United States' first national park, being also the world's first national park. In some European and Asian countries, however, national protection and nature reserves already existed, such as a part of the Forest of Fontainebleau.

Yellowstone was part of a federally governed territory. With no state government that could assume stewardship of the land, the federal government took on direct responsibility for the park, the official first national park of the United States. The combined effort and interest of conservationists, politicians and the Northern Pacific Railroad ensured the passage of enabling legislation by the United States Congress to create Yellowstone National Park. Theodore Roosevelt and his group of conservationists, the Boone and Crockett Club, were already an active campaigners, and so influential, as good stump speakers were highly necessary in the pre-telecommunications era, was highly influential in convincing fellow Republicans and big business to back the bill. Yellowstone National Park soon played a pivotal role in the conservation of these national treasures, as it was suffering at the hands of poachers and others who stood at the ready to pillage what they could from the area. Theodore Roosevelt and his newly formed Boone and Crockett Club successfully took the lead in protecting Yellowstone National Park from this plight, resulting in laws designed to conserve the natural resources in Yellowstone and other parks under the Government's purview.

American Pulitzer Prize-winning author Wallace Stegner wrote:

> "National parks are the best idea we ever had. Absolutely American, absolutely democratic, they reflect us at our best rather than our worst".

In his book Dispossessing the Wilderness: Indian Removal and the Making of the National Parks, Mark David Spence made the point that in order to create these uninhabited spaces, the United States first had to dispossess the Indians who were living in them.

International Growth of National Parks

Royal National Park.

The first area to use "national park" in its creation legislation was the U.S.'s Mackinac National Park, in 1875. (The area was later transferred to the state's authority in 1895, thus losing its official "national park" status).

Following the idea established in Yellowstone and Mackinac, there soon followed parks in other nations. In Australia, the Royal National Park was established just south of Sydney on 26 April 1879, becoming the world's second official national park. Since Mackinac lost its national park status, Royal National Park is, by some considerations, the second oldest national park now in existence.

Rocky Mountain National Park became Canada's first national park in 1885. New Zealand established Tongariro National Park in 1887.

In Europe, the first national parks were a set of nine parks in Sweden in 1909, followed by the Swiss National Park in 1914. Africa's first national park was established in 1925 when Albert I of Belgium designated an area of what is now Democratic Republic of Congo centred on the Virunga Mountains as the Albert National Park (since renamed

Virunga National Park). In 1926, the government of South Africa designated Kruger National Park as the nation's first national park, although it was an expansion of the earlier Sabie Game Reserve established in 1898 by President Paul Kruger of the old South African Republic, after whom the park was named. Argentina became the third country in the Americas to create a national park system, with the creation of the Nahuel Huapi National Park in 1934, through the initiative of Francisco Moreno.

After World War II, national parks were founded all over the world. The United Kingdom designated its first national park, Peak District National Park, in 1951. This followed perhaps 70 years of pressure for greater public access to the landscape. By the end of the decade a further nine national parks had been designated in the UK. Europe has some 359 national parks as of 2010. The Vanoise National Park in the Alps was the first French national park, created in 1963 after public mobilization against a touristic project.

In 1971, Lahemaa National Park in Estonia was the first area to be designated a national park in the former Soviet Union. In 1973, Mount Kilimanjaro was classified as a National Park and was opened to public access in 1977. In 1989, the Qomolangma National Nature Preserve (QNNP) was created to protect 3.381 million hectares on the north slope of Mount Everest in the Tibet Autonomous Region of China. This national park is the first major global park to have no separate warden and protection staff—all of its management being done through existing local authorities, allowing a lower cost basis and a larger geographical coverage (in 1989 when created, it was the largest protected area in Asia). It includes four of the six highest mountains Everest, Lhotse, Makalu, and Cho Oyu. The QNNP is contiguous to four Nepali national parks, creating a transborder conservation area equal in size to Switzerland.

National Parks Services

The world's first national park service was established in Canada. The Dominion Forest Reserves and Parks Act placed the dominion parks under the administration of the Dominion Park Branch (now Parks Canada). The branch was established to "protect sites of natural wonder" to provide a recreational experience, centered on the idea of the natural world providing rest and spiritual renewal from the urban setting. Canada now has the largest protected area in the world with 450,000 km² of national park space.

Even with the creation of Yellowstone, Yosemite, and nearly 37 other national parks and monuments, another 44 years passed before an agency was created in the United States to administer these units in a comprehensive way – the U.S. National Park Service (NPS). The 64th United States Congress passed the National Park Service Organic Act, which President Woodrow Wilson signed into law. Of the 419 sites managed by the National Park Service of the United States, only 61 carry the designation of National Park.

Notable Parks

The largest national park in the world meeting the IUCN definition is the Northeast Greenland National Park, which was established in 1974.

The smallest official national park in the world is Isles des Madeleines National Park. Its area of just.45 square kilometres (0.17 sq mi) was established as a national park in 1976.

Economic Ramifications

Countries with a large ecotourism industry, such as Costa Rica, often experience a huge economic effect on park management as well as the economy of the country as a whole.

Tourism

Tourism to national parks has increased considerably over time. In Costa Rica for example, a megadiverse country, tourism to parks has increased by 400% from 1985 to 1999. The term *national park* is perceived as a brand name that is associated with nature-based tourism and it symbolizes a "high quality natural environment with a well-designed tourist infrastructure".

Staff

The duties of a park ranger are to supervise, manage, and perform work in the conservation and use of Federal park resources. This involves functions such as park conservation; natural, historical, and cultural resource management; and the development and operation of interpretive and recreational programs for the benefit of the visiting public. Park rangers also have fire fighting responsibilities and execute search and rescue missions. Activities also include heritage interpretation to disseminate information to visitors of general, historical, or scientific information. Management of resources such as wildlife, lakeshores, seashores, forests, historic buildings, battlefields, archeological properties, and recreation areas are also part of the job of a park ranger. Since the establishment of the National Park Service in the US in 1916, the role of the park ranger has shifted from merely being a custodian of natural resources to include several activities that are associated with law enforcement. They control traffic and investigate violations, complaints, trespass/encroachment, and accidents.

Northeast Greenland National Park

Northeast Greenland National Park is the world's largest national park and the 9th largest protected area (the only larger protected areas all consist mostly of sea). Established in 1974 and expanded to its present size in 1988, it protects 972,000 km^2 (375,000 sq mi) of the interior and northeastern coast of Greenland and is bigger than all but 29 of the world's 196 countries. It was the first national park to be created in the Kingdom of

Denmark and remains Greenland's only national park. It is the northernmost national park in the world. It is the second largest by area of any second level subdivision of any country in the world trailing only the Qikiqtaaluk Region, Nunavut, Canada.

Geography

The park shares borders, largely laid out as straight lines, with the Sermersooq municipality in the south and with the Avannaata municipality in the west along the 45° West meridian on the ice cap in the west. The large interior of the park is part of the Greenland Ice Sheet, but there are also large ice-free areas along the coast and on Peary Land in the north. The park includes the King Frederick VIII Land and King Christian X Land geographical areas.

The area is subject to larger loss of ice than expected.

Population

Zackenberg Station.

The park has no permanent human population, although 400 sites see occasional summertime use. In 1986, the population of the park was 40, living at Mestersvig. These 40 were involved in cleanup and closeout operations at mining exploration sites and soon left. Since then censuses have recorded zero permanent human population. Recently only 31 people and about 110 dogs were present over winter in North East Greenland, distributed among the following stations (all on the coast, except Summit Camp):

- Daneborg: Headquarters of the Sirius Patrol, the park policing agency.
- Danmarkshavn: Civilian weather station.
- Station Nord: Military base.
- Mestersvig: Military outpost with 1,800 m gravel runway.
- Zackenberg: Summer-only research station.
- Summit Camp: Research station on the Greenland Ice Sheet.

During summer scientists add to these numbers. The research station ZERO (Zackenberg Ecological Research Operations) 74°28′11″N 20°34′15″W can cater for over 20 scientists and station personnel.

Fauna

Iceberg in the national park.

An estimated 5,000 to 15,000 musk oxen, as well as numerous polar bears and walrus, can be found near the coastal regions of the park. In 1993, this was estimated to be 40% of the world population of musk ox. Other mammals include Arctic fox, stoat, collared lemming, Arctic hare and a small but important population of Greenland wolf. Other marine mammals include ringed seal, bearded seal, harp seal and hooded seal as well as narwhal and beluga whale.

Species of birds which breed in the park include great northern diver, barnacle goose, pink-footed goose, common eider, king eider, gyrfalcon, snowy owl, sanderling, ptarmigan and raven.

Death Valley National Park

Sand dunes in Death Valley National Park.

Death Valley National Park is an American national park that straddles the California–Nevada border, east of the Sierra Nevada. The park boundaries include Death Valley, the northern section of Panamint Valley, the southern section of Eureka Valley, and most of

Saline Valley. The park occupies an interface zone between the arid Great Basin and Mojave deserts, protecting the northwest corner of the Mojave Desert and its diverse environment of salt-flats, sand dunes, badlands, valleys, canyons, and mountains. Death Valley is the largest national park in the contiguous United States, and the hottest, driest and lowest of all the national parks in the United States. The second-lowest point in the Western Hemisphere is in Badwater Basin, which is 282 feet (86 m) below sea level. Approximately 91% of the park is a designated wilderness area. The park is home to many species of plants and animals that have adapted to this harsh desert environment. Some examples include creosote bush, bighorn sheep, coyote, and the Death Valley pupfish, a survivor from much wetter times. UNESCO included Death Valley as the principal feature of its Mojave and Colorado Deserts Biosphere Reserve in 1984.

A series of Native American groups inhabited the area from as early as 7000 BC, most recently the Timbisha around 1000 AD who migrated between winter camps in the valleys and summer grounds in the mountains. A group of European-Americans, trapped in the valley in 1849 while looking for a shortcut to the gold fields of California, gave the valley its name, even though only one of their group died there. Several short-lived boom towns sprang up during the late 19th and early 20th centuries to mine gold and silver. The only long-term profitable ore to be mined was borax, which was transported out of the valley with twenty-mule teams. The valley later became the subject of books, radio programs, television series, and movies. Tourism expanded in the 1920s when resorts were built around Stovepipe Wells and Furnace Creek. Death Valley National Monument was declared in 1933 and the park was substantially expanded and became a national park in 1994.

The natural environment of the area has been shaped largely by its geology. The valley is actually a graben with the oldest rocks being extensively metamorphosed and at least 1.7 billion years old. Ancient, warm, shallow seas deposited marine sediments until rifting opened the Pacific Ocean. Additional sedimentation occurred until a subduction zone formed off the coast. The subduction uplifted the region out of the sea and created a line of volcanoes. Later the crust started to pull apart, creating the current Basin and Range landform. Valleys filled with sediment and, during the wet times of glacial periods, with lakes, such as Lake Manly.

In 2013, Death Valley National Park was designated as a dark sky park by the International Dark-Sky Association.

Geographic Setting

There are two major valleys in the park, Death Valley and Panamint Valley. Both of these valleys were formed within the last few million years and both are bounded by north–south-trending mountain ranges. These and adjacent valleys follow the general trend of Basin and Range topography with one modification: there are parallel strike-slip faults that perpendicularly bound the central extent of Death Valley. The result of this shearing action is additional extension in the central part of Death Valley which causes a slight widening and more subsidence there.

Uplift of surrounding mountain ranges and subsidence of the valley floor are both occurring. The uplift on the Black Mountains is so fast that the alluvial fans (fan-shaped deposits at the mouth of canyons) there are small and steep compared to the huge alluvial fans coming off the Panamint Range. Fast uplift of a mountain range in an arid environment often does not allow its canyons enough time to cut a classic V-shape all the way down to the stream bed. Instead, a V-shape ends at a slot canyon halfway down, forming a 'wine glass canyon.' Sediment is deposited on a small and steep alluvial fan.

At 282 feet (86 m) below sea level at its lowest point, Badwater Basin on Death Valley's floor is the second-lowest depression in the Western Hemisphere (behind Laguna del Carbón in Argentina), while Mount Whitney, only 85 miles (137 km) to the west, rises to 14,505 feet (4,421 m). This topographic relief is the greatest elevation gradient in the contiguous United States and is the terminus point of the Great Basin's southwestern drainage. Although the extreme lack of water in the Great Basin makes this distinction of little current practical use, it does mean that in wetter times the lake that once filled Death Valley (Lake Manly) was the last stop for water flowing in the region, meaning the water there was saturated in dissolved materials. Thus the salt pans in Death Valley are among the largest in the world and are rich in minerals, such as borax and various salts and hydrates. The largest salt pan in the park extends 40 miles (64 km) from the Ashford Mill Site to the Salt Creek Hills, covering some 200 square miles (520 km^2) of the valley floor. The best known playa in the park is the Racetrack, known for its moving rocks.

Climate

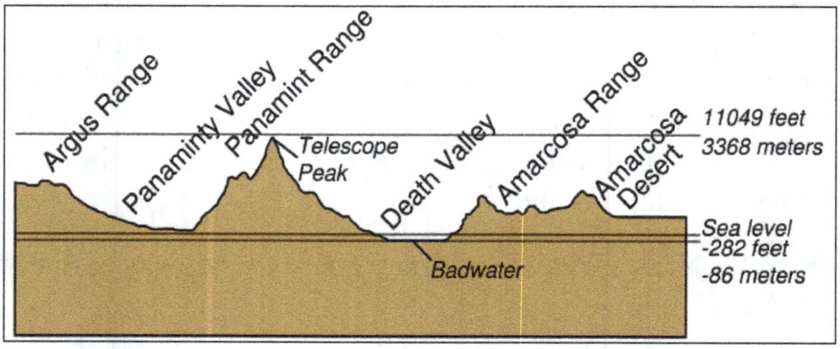

A cross section through the highest and lowest points in Death Valley National Park.

According to the Köppen climate classification system, Death Valley National Park has a Hot Desert Climate (*DWh*). The plant hardiness zone at Badwater Basin is 9b with an average annual extreme minimum temperature of 27.3 °F (-2.6 °C).

Death Valley is the hottest and driest place in North America due to its lack of surface water and low relief. It is so frequently the hottest spot in the United States that many tabulations of the highest daily temperatures in the country omit Death Valley as a matter of course.

On the afternoon of July 10, 1913, the United States Weather Bureau recorded a high temperature of 134 °F (56.7 °C) at Greenland Ranch (now Furnace Creek) in Death Valley. This temperature stands as the highest ambient air temperature ever recorded at the surface of the Earth. (A report of a temperature of 58 °C (136.4 °F) recorded in Libya in 1922 was later determined to be inaccurate). Daily summer temperatures of 120 °F (49 °C) or greater are common, as well as below freezing nightly temperatures in the winter. July is the hottest month, with an average high of 115 °F (46 °C) and an average low of 88 °F (31 °C). December is the coldest month, with an average high of 65 °F (18 °C) and an average low of 39 °F (4 °C). The record low is 15 °F (−9.4 °C).

Several of the larger Death Valley springs derive their water from a regional aquifer, which extends as far east as southern Nevada and Utah. Much of the water in this aquifer has been there for many thousands of years, since the Pleistocene ice ages, when the climate was cooler and wetter. Today's drier climate does not provide enough precipitation to recharge the aquifer at the rate at which water is being withdrawn.

Telescope Peak.

The highest range within the park is the Panamint Range with Telescope Peak being its highest point at 11,049 feet (3,368 m). The Death Valley region is a transitional zone in the northernmost part of the Mojave Desert and consists of five mountain ranges removed from the Pacific Ocean. Three of these are significant barriers: the Sierra Nevada, the Argus Range, and the Panamint Range. Air masses tend to lose moisture as they are forced up over mountain ranges, in what climatologists call a rainshadow effect.

The exaggerated rainshadow effect for the Death Valley area makes it North America's driest spot, receiving about 1.5 inches (38 mm) of rainfall annually at Badwater (some years fail to register any measurable rainfall). Annual average precipitation varies from 1.92 inches (49 mm) overall below sea level to over 15 inches (380 mm) in the higher mountains that surround the valley. When rain does arrive it often does so in intense storms that cause flash floods which remodel the landscape and sometimes create very shallow ephemeral lakes.

Lake Badwater.

The hot, dry climate makes it difficult for soil to form. Mass wasting, the down-slope movement of loose rock, is therefore the dominant erosive force in mountainous area, resulting in "skeletonized" ranges (mountains with very little soil on them). Sand dunes in the park, while famous, are not nearly as widespread as their fame or the dryness of the area may suggest. The Mesquite Flat dune field is the most easily accessible from the paved road just east of Stovepipe Wells in the north-central part of the valley and is primarily made of quartz sand. Another dune field is just 10 miles (16 km) to the north but is instead mostly composed of travertine sand. The highest dunes in the park, and some of the highest in North America, are located in the Eureka Valley about 50 miles (80 km) to the north of Stovepipe Wells, while the Panamint Valley dunes and the Saline Valley dunes are located west and northwest of the town, respectively. The Ibex dune field is near the seldom-visited Ibex Hill in the southernmost part of the park, just south of the Saratoga Springs marshland. All of the latter four dune fields are accessible only via unpaved roads. Prevailing winds in the winter come from the north, and prevailing winds in the summer come from the south. Thus the overall position of the dune fields remains more or less fixed.

There are rare exceptions to the dry nature of the area. In 2005, an unusually wet winter created a 'lake' in the Badwater Basin and led to the greatest wildflower season in the park's history. In October 2015, a "1000 year flood event" with over three inches of rain caused major damage in Death Valley National Park.

Geologic

Era	Rock Units/Formations	Principal Geologic Events
Cenozoic	Alluvial fans, stream, and playa deposits, dunes, numerous sedimentary, volcanic, and plutonic units in separate and interconnected basins and igneous fields (includes Artist Drive, Furnace Creek, Funeral, and Nova Formations).	Major unconformity, continued deposition in modern Death Valley, opening of modern Death Valley, continuing development of present ranges and basins, onset of major extension.
Mesozoic	Granitic plutons, Butte Valley.	Thrust faulting and intrusion of plutons related to Sierra Nevada batholith; shallow marine deposition; unconformity.

Paleozoic	Resting spring Shale, Tin Mountain Limestone, Lost Burro, Hidden Valley Dolomite, Eureka Quartzite, Nopah, Bonanza King, Carrara, Zabriskie Quartzite, Wood Canyon.	Development of a long-continuing carbonate bank on a passive continental margin; numerous intervals of emergence, interrupted by deposition of a blanket of sandstone in Middle Ordovician time. Deposition of a wedge of silliciclastic sediment during and immediately following the rifting along a new continental margin.
Proterozoic	Crystalline basement, Pahrump, Stirling Quartzite, Johnnie, Ibex, Noonday Dolomite, Kingston Peak, Beck Spring, Crystal Spring.	Regional metamorphism, Major unconformity, rapid uplift and erosion, shallow marine deposition, glacio-marine deposition, unconformity. Shallow to deep marine deposition along an incipient continental margin.

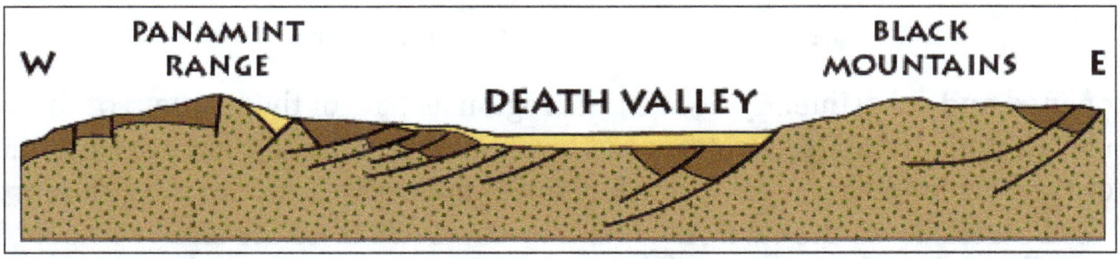

The Death Valley basin is filled with sediment (light yellow) eroded from the surrounding mountains. Black lines show some of the major faults that created the valley.

The park has a diverse and complex geologic history. Since its formation, the area that comprises the park has experienced at least four major periods of extensive volcanism, three or four periods of major sedimentation, and several intervals of major tectonic deformation where the crust has been reshaped. Two periods of glaciation (a series of ice ages) have also had effects on the area, although no glaciers ever existed in the ranges now in the park.

Basement and Pahrump Group

Little is known about the history of the oldest exposed rocks in the area due to extensive metamorphism (alteration of rock by heat and pressure). Radiometric dating gives an age of 1,700 million years for the metamorphism during the Proterozoic. About 1,400 million years ago a mass of granite now in the Panamint Range intruded this complex. Uplift later exposed these rocks to nearly 500 million years of erosion.

The Proterozoic sedimentary formations of the Pahrump Group were deposited on these basement rocks. This occurred following uplift and erosion of any earlier sediments from the Proterozoic basement rocks. The Pahrump is composed of arkose conglomerate (quartz clasts in a concrete-like matrix) and mudstone in its lower part, followed by dolomite from carbonate banks topped by algal mats as stromatolites, and finished with basin-filling sediment derived from the above, including possible glacial till from the hypothesized Snowball Earth glaciation. The very youngest rocks in the Pahrump Group are basaltic lava flows.

Rifting and Deposition

The Noonday Dolomite was formed as a carbonate shelf after the break-up of Rodinia.

A rift opened and subsequently flooded the region as part of the breakup of the supercontinent Rodinia in the Neoproterozoic (by about 755 million years ago) and the creation of the Pacific Ocean. A shoreline similar to the present Atlantic Ocean margin of the United States lay to the east. An algal mat-covered carbonate bank was deposited, forming the Noonday Dolomite. Subsidence of the region occurred as the continental crust thinned and the newly formed Pacific widened, forming the Ibex Formation. An angular unconformity (an uneven gap in the geologic record) followed.

A true ocean basin developed to the west, breaking all the earlier formations along a steep front. A wedge of clastic sediment then began to accumulate at the base of the two underwater precipices, starting the formation of opposing continental shelves. Three formations developed from sediment that accumulated on the wedge. The region's first known fossils of complex life are found in the resulting formations. Notable among these are the Ediacara fauna and trilobites, the evolution of the latter being part of the Cambrian Explosion of life.

The sandy mudflats gave way about 550 million years ago to a carbonate platform (similar to the one around the present-day Bahamas), which lasted for the next 300 million years of Paleozoic time. Death Valley's position was then within ten or twenty degrees of the Paleozoic equator. Thick beds of carbonate-rich sediments were periodically interrupted by periods of emergence. Although details of geography varied during this immense interval of time, a north-northeasterly trending coastline generally ran from Arizona up through Utah. The resulting eight formations and one group are 20,000 feet (6 km) thick and underlay much of the Cottonwood, Funeral, Grapevine, and Panamint ranges.

Compression and Uplift

In the early-to-mid- Mesozoic the western edge of the North American continent was pushed against the oceanic plate under the Pacific Ocean, creating a subduction zone.

A subduction zone is a type of contact between different crustal plates where heavier crust slides below lighter crust. Erupting volcanoes and uplifting mountains were created as a result, and the coastline was pushed to the west. The Sierran Arc started to form to the northwest from heat and pressure generated from subduction, and compressive forces caused thrust faults to develop.

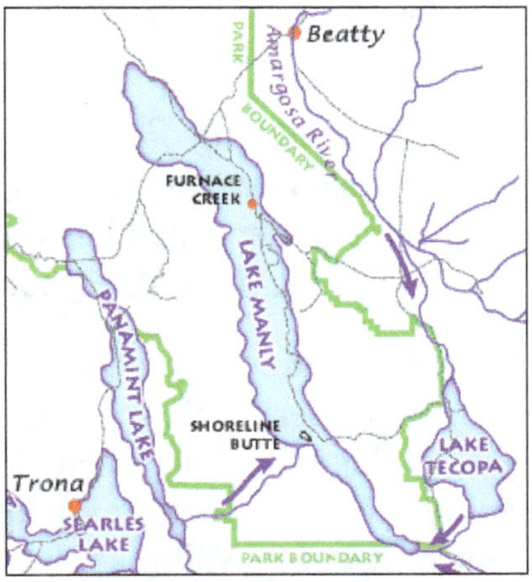

The Lake Manly lake system as it might have looked during its last maximum extent 22,000 years ago.

A long period of uplift and erosion was concurrent with and followed the above events, creating a major unconformity, which is a large gap in the geologic record. Sediments worn off the Death Valley region were carried both east and west by wind and water. No Jurassic- to Eocene-aged sedimentary formations exist in the area, except for some possibly Jurassic-age volcanic rocks.

Stretching and Lakes

During very wet periods, the Amargosa River can flow at the surface.

Basin and Range-associated stretching of large parts of crust below southwestern United States and northwestern Mexico started around 16 million years ago and the region is still spreading. This stretching began to affect the Death and Panamint valleys area by 3 million years ago. Before this, rocks now in the Panamint Range were on top of rocks that would become the Black Mountains and the Cottonwood Mountains. Lateral and vertical transport of these blocks was accomplished by movement on normal faults. Right-lateral movement along strike-slip faults that run parallel to and at the base of the ranges also helped to develop the area. Torsional forces, probably associated with northwesterly movement of the Pacific Plate along the San Andreas Fault (west of the region), is responsible for the lateral movement.

Igneous activity associated with this stretching occurred from 12 million to 4 million years ago. Sedimentation is concentrated in valleys (basins) from material eroded from adjacent ranges. The amount of sediment deposited has roughly kept up with this subsidence, resulting in retention of more or less the same valley floor elevation over time.

Pleistocene ice ages started 2 million years ago, and melt from alpine glaciers on the nearby Sierra Nevada Mountains fed a series of lakes that filled Death and Panamint valleys and surrounding basins. The lake that filled Death Valley was the last of a chain of lakes fed by the Amargosa and Mojave Rivers, and possibly also the Owens River. The large lake that covered much of Death Valley's floor, which geologists call Lake Manly, started to dry up 10,500 years ago. Saltpans and playas were created as ice age glaciers retreated, thus drastically reducing the lakes' water source. Only faint shorelines are left.

Biology

Sphinx moth on a rock nettle in Mosaic Canyon.

Habitat varies from saltpan at 282 feet (86 m) below sea level to the sub-alpine conditions found on the summit of Telescope Peak, which rises to 11,049 feet (3,368 m). Vegetation zones include creosote bush, desert holly, and mesquite at the lower elevations and sage up through shadscale, blackbrush, Joshua tree, pinyon-juniper, to limber pine and bristlecone pine woodlands. The saltpan is devoid of vegetation, and the rest of the valley floor and lower slopes have sparse cover, although where water is available, an abundance of vegetation is usually present. These zones and the adjacent desert support a variety of wildlife species, including 51 species of native mammals, 307 species of birds, 36 species of reptiles, 3 species of amphibians, and 2 species of native fish.

Small mammals are more numerous than large mammals, such as bighorn sheep, coyotes, bobcats, kit foxes, cougars, and mule deer. Mule deer are present in the pinyon/juniper associations of the Grapevine, Cottonwood, and Panamint ranges. Bighorn sheep are a rare species of mountain-dwelling sheep that exist in isolated bands in the Sierra and in Death Valley. These are highly adaptable animals and can eat almost any plant. They have no known predators, but humans and burros compete for habitat.

Death Valley pupfish spawning in Salt Creek.

The ancestors of the Death Valley pupfish swam to the area from the Colorado River via a long-since dried-up system of rivers and lakes. They now live in two separate populations: One in Salt Creek and another in Cottonball Marsh. Death Valley is one of the hottest and driest places in North America, yet it is home to over 1,000 species of plants; 23 of which, including the very rare rock lady (Holmgrenanthe petrophila), are not found anywhere else.

Adaptation to the dry environment is key. For example, creosote bush and mesquite have tap-root systems that can extend 50 feet (15 m) down in order to take advantage of a year-round supply of ground water. The diversity of Death Valley's plant communities results partly from the region's location in a transition zone between the Mojave Desert, the Great Basin Desert and the Sonoran Desert. This location, combined with the great relief found within the park, supports vegetation typical of three biotic life zones: the lower Sonoran, the Canadian, and the arctic/alpine in portions of the Panamint Range. Based on the Munz and Keck classifications, seven plant communities can be categorized within these life zones, each characterized by dominant vegetation and representative of three vegetation types: scrub, desert woodland, and coniferous forest. Microhabitats further subdivide some communities into zones, especially on the valley floor.

Unlike more typical locations across the Mojave Desert, many of the water-dependent Death Valley habitats possess a diversity of plant and animal species that are not found anywhere else in the world. The existence of these species is due largely to a unique geologic history and the process of evolution that has progressed in habitats that have been isolated from one another since the Pleistocene epoch.

Activities

Sightseeing is available by personal automobile, four-wheel drive, bicycle, mountain bike (on established roadways only), and hiking. Riding through the park on motorcycle is also a popular pastime. State Route 190, the Badwater Road, the Scotty's Castle Road, and paved roads to Dante's View and Wildrose provide access to the major scenic viewpoints and historic points of interest. More than 350 miles (560 km) of unpaved and four-wheel-drive roads provide access to wilderness hiking, camping, and historical sites. All vehicles must be licensed and street legal. There are hiking trails of varying lengths and difficulties, but most backcountry areas are accessible only by cross-country hiking. There are thousands of hiking possibilities. The normal season for visiting the park is from October 15 to May 15, avoiding summer extremes in temperature. Costumed living history tours of the historic Death Valley Scotty's Castle were conducted for a fee, but were suspended in October 2015 due to flood damage to the buildings and grounds. It is not expected to re-open until 2020.

A tourist sliding down Star Dune in the Mesquite Flat dune field.

There are nine designated campgrounds within the park, and overnight backcountry camping permits are available at the Visitor Center. Xanterra Parks and Resorts owns and operates a private resort, the Oasis at Death Valley, which comprises two separate and distinct hotels: the Inn at Death Valley is a four-star historic hotel, and the Ranch at Death Valley is a three-star ranch-style property reminiscent of the mining and prospecting days. Panamint Springs Resort is in the western part of the park. Death Valley Lodging Company operates the Stovepipe Wells Resort under a concession permit. There are a few motels near entrances to the park, in Shoshone, Death Valley Junction, Beatty, and Pahrump.

Furnace Creek Visitor Center is located on CA-190. A 22-minute introductory slide program is shown every 30 minutes. During the winter season—November through April—rangers offer interpretive tours and a wide variety of walks, talks, and slide presentations about Death Valley cultural and natural history. The visitor center has displays dealing with the park's geology, climate, wildlife and natural history. There are also specific sections dealing with the human history and pioneer experience. The Death Valley Natural History Association maintains a bookstore specifically geared to the natural and cultural history of the park.

The northeast corner of Saline Valley has several developed hot spring pools. The pools can be accessed by driving on the unpaved Saline Valley Road for several hours, or by flying a personal aircraft to the Chicken Strip—an uncharted airstrip a short walk from the springs.

Death Valley National Park is a popular location for stargazing as it has one of the darkest night skies in the United States. Despite its remote location, air quality and night visibility are threatened by civilization. In particular, light pollution is introduced by nearby Las Vegas. The darkest skies are, in general, located in the northwest of the park. The northwestern area of the park, including sites such as Ubehebe Crater, is a Bortle class 1 or "excellent dark sky" site. The Andromeda Galaxy and the Triangulum Galaxy are visible to the unaided eye under these conditions, and the Milky Way casts shadows; optical phenomenon such as zodiacal light or "false dawn" and gegenschein are also visible to the unaided eye under these conditions. Most southern regions of the park are Bortle class 2 or "average dark sky" sites.

A 360-degree panorama of Racetrack Playa at night. The Milky Way is visible as an arc in the center.

Wood Buffalo National Park

Wood Buffalo National Park is the largest National Park of Canada at 44,807 km^2 (17,300 sq mi). It is located in northeastern Alberta and the southern Northwest Territories. Larger in area than Switzerland, it is the second-largest national park in the world. The park was established in 1922 to protect the world's largest herd of free roaming wood bison, currently estimated at more than 5,000. It is one of two known nesting sites of whooping cranes.

The park ranges in elevation from 183 m (600 ft) at the Little Buffalo River to 945 m (3,100 ft) in the Caribou Mountains. The park headquarters is located in Fort Smith, with a smaller satellite office in Fort Chipewyan, Alberta. The park contains one of the world's largest fresh water deltas, the Peace-Athabasca Delta, formed by the Peace, Athabasca and Birch Rivers. It is also known for its karst sinkholes in the north-eastern section of the park. Alberta's largest springs (by volume, with an estimated discharge rate of eight cubic meters per second), Neon Lake Springs, are located in the Jackfish River drainage. Wood Buffalo is located directly north of the Athabasca Oil Sands.

This area was designated a UNESCO World Heritage Site in 1983 for the biological diversity of the Peace-Athabasca Delta, one of the world's largest freshwater deltas, as well as the population of wild bison. The Royal Astronomical Society of Canada designated Wood Buffalo National Park as Canada's newest and the world's largest dark-sky preserve. The designation helps preserve nighttime ecology for the park's large populations of bats, night hawks and owls, as well as providing opportunities for visitors to experience the northern lights.

As a National Park

Established in 1922, the park was created on Crown land acquired the territory of Treaty 8 between Canada and the local First Nations. The park itself completely surrounds several Indian reserves such as Peace Point.

Between 1925 and 1928, over 6,000 plains bison were introduced to the park, where they hybridized with the local wood bison, as well as introducing bovine tuberculosis and brucellosis diseases into the herd. Parks officials have since that time attempted to undo this damage with successive culls of diseased animals. In 1957, however, a disease-free, wood bison herd of 200 was discovered near Nyarling river in Wood Buffalo National Park. In 1965, 23 of these bison were relocated to the south side of Elk Island National Park and 300 remain there today as the most genetically pure wood bison remaining. Between 1951 and 1967, 4000 bison were killed and 2,000,000 pounds (910 t) of meat were sold from a special abattoir built at Hay Camp. These smaller culls did not eradicate the diseases, however, and in 1990 a plan was announced to cull the entire herd and restock it with undiseased animals from Elk Island National Park. This plan was abandoned due to a negative public reaction to the announcement. Since that time, wolves, the bison's main predator, have recovered in numbers due to a reduction in control efforts (mostly poisoning), reducing the size of the herd.

In 1983, a 21-year lease was granted to Canadian Forest Products Ltd. to log a 50,000-hectare area of Wood Buffalo National Park. The Canadian Parks and Wilderness Society filed a lawsuit against Parks Canada for violating the National Parks Act. Before the trial commenced in 1992, Parks Canada acquiesced and recognized that the lease was invalid and unauthorized by the provisions of the act.

In March 2019, a new provincial park, known as Kitaskino Nuwenëné Wildland Provincial Park, was established on the borders of the Wood Buffalo National Park. The protection of this park was first proposed by the Mikisew Cree First Nation and it will protect the natural ecosystems from the expanding industrial areas north of Fort McMurray. The park was created after three oil companies – Teck Resources, Cenovus Energy, and Imperial Oil – voluntarily gave up certain oilsands and mining leases in the area, following negotiations with the Alberta government and indigenous groups. This new provincial park will be closed to forestry and new energy projects, but existing wells in the area can keep producing and traditional indigenous land uses are allowed.

Beaver Lodge.

Climate

In the park, summers are very short, but days are long. Temperatures range between 10 to 30 °C (50.0 to 86.0 °F) during this season. On average, summers are characterized by warm and dry days although in some years, it can have cool and wet days. The mean high in July is 22.5 °C (72.5 °F) while the mean low is 9.5 °C (49.1 °F). Fall tends to have cool, windy and dry days in which the first snowfall usually occurs in October. Winters are cold with temperatures that can drop below −40 °C (−40.0 °F) in January and February, the coldest months. The mean high in January is −21.7 °C (−7.1 °F) while the mean low is −31.8 °C (−25.2 °F). In spring, temperatures gradually warm up as the days become longer.

Wildlife

Wood Bison.

Wood Buffalo National Park contains a large variety of wildlife species, such as moose, bison, great grey owls, black bears, hawks, spotted owls, timber wolves, lynxes, beavers, snowy owls, marmots, bald eagles, martens, wolverines, peregrine falcons, whooping cranes, snowshoe hares, sandhill cranes, ruffed grouses, and the world's northernmost population of red-sided garter snakes, which form communal dens within the park.

Wood Buffalo Park contains the only natural nesting habitat for the endangered whooping crane. Known as Whooping Crane Summer Range, it is classified as a Ramsar site.

It was identified through the International Biological Program. The range is a complex of contiguous water bodies, primarily lakes and various wetlands, such as marshes and bogs, but also includes streams and ponds.

In 2007, the world's largest beaver dam – about 850-metre (2,790 ft) in length – was discovered in the park using satellite imagery; The dam, located at 58°16.3′N 112°15.1′W58.2717°N 112.2517°W, about 200 kilometres (120 mi) from Fort Chipewyan, had only been sighted by satellite and fixed-wing aircraft until July 2014.

Transportation

Year-round access is available to Fort Smith by road on the Mackenzie Highway, which connects to Highway 5 near Hay River, Northwest Territories. Commercial flights are available to Fort Smith and Fort Chipewyan from Edmonton. Winter access is also available using winter and ice roads from Fort McMurray through Fort Chipewyan.

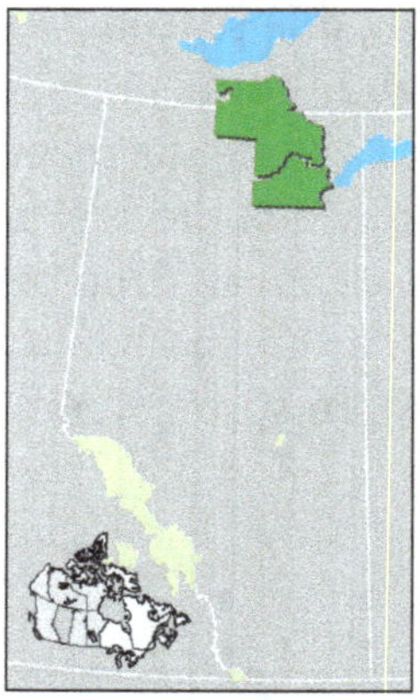

Location and extent.

Yellowstone National Park

Yellowstone National Park is an American national park located mostly in Wyoming, with small sections in Montana and Idaho. It was established by the U.S. Congress and signed into law by President Ulysses S. Grant. Yellowstone was the first national park in the U.S. and is also widely held to be the first national park in the world. The park is known for its wildlife and its many geothermal features, especially Old Faithful geyser, one of its most popular features. It has many types of ecosystem, but the subalpine forest is the most abundant. It is part of the South Central Rockies forests ecoregion.

Native Americans have lived in the Yellowstone region for at least 11,000 years. Aside from visits by mountain men during the early-to-mid-19th century, organized exploration did not begin until the late 1860s. Management and control of the park originally fell under the jurisdiction of the Secretary of the Interior, the first being Columbus Delano. However, the U.S. Army was subsequently commissioned to oversee management of Yellowstone for a 30-year period between 1886 and 1916. In 1917, administration of the park was transferred to the National Park Service, which had been created the previous year. Hundreds of structures have been built and are protected for their architectural and historical significance, and researchers have examined more than a thousand archaeological sites.

Yellowstone National Park spans an area of 3,468.4 square miles (8,983 km^2), comprising lakes, canyons, rivers and mountain ranges. Yellowstone Lake is one of the largest high-elevation lakes in North America and is centered over the Yellowstone Caldera, the largest supervolcano on the continent. The caldera is considered a dormant volcano. It has erupted with tremendous force several times in the last two million years. Half of the world's geysers and hydrothermal features are in Yellowstone, fueled by this ongoing volcanism. Lava flows and rocks from volcanic eruptions cover most of the land area of Yellowstone. The park is the centerpiece of the Greater Yellowstone Ecosystem, the largest remaining nearly-intact ecosystem in the Earth's northern temperate zone. In 1978, Yellowstone was named a UNESCO World Heritage Site.

Hundreds of species of mammals, birds, fish, and reptiles have been documented, including several that are either endangered or threatened. The vast forests and grasslands also include unique species of plants. Yellowstone Park is the largest and most famous megafauna location in the contiguous United States. Grizzly bears, wolves, and free-ranging herds of bison and elk live in this park. The Yellowstone Park bison herd is the oldest and largest public bison herd in the United States. Forest fires occur in the park each year; in the large forest fires of 1988, nearly one third of the park was burnt. Yellowstone has numerous recreational opportunities, including hiking, camping, boating, fishing and sightseeing. Paved roads provide close access to the major geothermal areas as well as some of the lakes and waterfalls. During the winter, visitors often access the park by way of guided tours that use either snow coaches or snowmobiles.

Detailed pictorial map.

The park contains the headwaters of the Yellowstone River, from which it takes its historical name. Near the end of the 18th century, French trappers named the river *Roche Jaune*, which is probably a translation of the Hidatsa name *Mi tsi a-da-zi* ("Yellow Rock River"). Later, American trappers rendered the French name in English as "Yellow Stone". Although it is commonly believed that the river was named for the yellow rocks seen in the Grand Canyon of the Yellowstone, the Native American name source is unclear.

The human history of the park began at least 11,000 years ago when Native Americans began to hunt and fish in the region. During the construction of the post office in Gardiner, Montana, in the 1950s, an obsidian projectile point of Clovis origin was found that dated from approximately 11,000 years ago. These Paleo-Indians, of the Clovis culture, used the significant amounts of obsidian found in the park to make cutting tools and weapons. Arrowheads made of Yellowstone obsidian have been found as far away as the Mississippi Valley, indicating that a regular obsidian trade existed between local tribes and tribes farther east. By the time white explorers first entered the region during the Lewis and Clark Expedition in 1805, they encountered the Nez Perce, Crow, and Shoshone tribes. While passing through present day Montana, the expedition members heard of the Yellowstone region to the south, but they did not investigate it.

In 1806, John Colter, a member of the Lewis and Clark Expedition, left to join a group of fur trappers. After splitting up with the other trappers in 1807, Colter passed through a portion of what later became the park, during the winter of 1807–1808. He observed at least one geothermal area in the northeastern section of the park, near Tower Fall. After surviving wounds he suffered in a battle with members of the Crow and Blackfoot tribes in 1809, Colter described a place of "fire and brimstone" that most people dismissed as delirium; the supposedly imaginary place was nicknamed "Colter's Hell". Over the next 40 years, numerous reports from mountain men and trappers told of boiling mud, steaming rivers, and petrified trees, yet most of these reports were believed at the time to be myth.

After an 1856 exploration, mountain man Jim Bridger (also believed to be the first or second European American to have seen the Great Salt Lake) reported observing boiling springs, spouting water, and a mountain of glass and yellow rock. These reports were largely ignored because Bridger was a known "spinner of yarns". In 1859, a U.S. Army Surveyor named Captain William F. Raynolds embarked on a two-year survey of the northern Rockies. After wintering in Wyoming, in May 1860, Raynolds and his party—which included naturalist Ferdinand Vandeveer Hayden and guide Jim Bridger—attempted to cross the Continental Divide over Two Ocean Plateau from the Wind River drainage in northwest Wyoming. Heavy spring snows prevented their passage, but had they been able to traverse the divide, the party would have been the first organized survey to enter the Yellowstone region. The American Civil War hampered further organized explorations until the late 1860s.

The first detailed expedition to the Yellowstone area was the Cook–Folsom–Peterson Expedition of 1869, which consisted of three privately funded explorers. The Folsom party followed the Yellowstone River to Yellowstone Lake. The members of the Folsom party kept a journal and based on the information it reported, a party of Montana residents organized the Washburn–Langford–Doane Expedition in 1870. It was headed by the surveyor-general of Montana Henry Washburn, and included Nathaniel

P. Langford (who later became known as "National Park" Langford) and a U.S. Army detachment commanded by Lt. Gustavus Doane.

The expedition spent about a month exploring the region, collecting specimens and naming sites of interest. A Montana writer and lawyer named Cornelius Hedges, who had been a member of the Washburn expedition, proposed that the region should be set aside and protected as a national park; he wrote detailed articles about his observations for the Helena Herald newspaper between 1870 and 1871. Hedges essentially restated comments made in October 1865 by acting Montana Territorial Governor Thomas Francis Meagher, who had previously commented that the region should be protected. Others made similar suggestions. In an 1871 letter from Jay Cooke to Ferdinand V. Hayden, Cooke wrote that his friend, Congressman William D. Kelley had also suggested "Congress pass a bill reserving the Great Geyser Basin as a public park forever".

Park Creation

In 1871, eleven years after his failed first effort, Ferdinand V. Hayden was finally able to explore the region. With government sponsorship, he returned to the region with a second, larger expedition, the Hayden Geological Survey of 1871. He compiled a comprehensive report, including large-format photographs by William Henry Jackson and paintings by Thomas Moran. The report helped to convince the U.S. Congress to withdraw this region from public auction. On March 1, 1872, President Ulysses S. Grant signed The Act of Dedication law that created Yellowstone National Park.

Hayden, while not the only person to have thought of creating a park in the region, was its first and most enthusiastic advocate. He believed in "setting aside the area as a pleasure ground for the benefit and enjoyment of the people" and warned that there were those who would come and "make merchandise of these beautiful specimens". Worrying the area could face the same fate as Niagara Falls, he concluded the site should "be as free as the air or Water." In his report to the Committee on Public Lands, he concluded that if the bill failed to become law, "the vandals who are now waiting to enter into this wonder-land, will in a single season despoil, beyond recovery, these remarkable curiosities, which have required all the cunning skill of nature thousands of years to prepare".

Hayden and his 1871 party recognized Yellowstone as a priceless treasure that would become rarer with time. He wished for others to see and experience it as well. Eventually the railroads and, sometime after that, the automobile would make that possible. The Park was not set aside strictly for ecological purposes; however, the designation "pleasure ground" was not an invitation to create an amusement park. Hayden imagined something akin to the scenic resorts and baths in England, Germany, and Switzerland:

> "The act of dedication: AN ACT to set apart a certain tract of land lying near the headwaters of the Yellowstone River as a public park. Be it enacted by the Senate and House of Representatives of the United States of America in Congress assembled, That the tract of land in the Territories of Montana and Wyoming is

> hereby reserved and withdrawn from settlement, occupancy, or sale under the laws of the United States, and dedicated and set apart as a public park or pleasuring ground for the benefit and enjoyment of the people; and all persons who shall locate, or settle upon, or occupy the same or any part thereof, except as hereinafter provided, shall be considered trespassers and removed there from."

There was considerable local opposition to the Yellowstone National Park during its early years. Some of the locals feared that the regional economy would be unable to thrive if there remained strict federal prohibitions against resource development or settlement within park boundaries and local entrepreneurs advocated reducing the size of the park so that mining, hunting, and logging activities could be developed. To this end, numerous bills were introduced into Congress by Montana representatives who sought to remove the federal land-use restrictions.

After the park's official formation, Nathaniel Langford was appointed as the park's first superintendent in 1872 by Secretary of Interior Columbus Delano, the first overseer and controller of the park. Langford served for five years but was denied a salary, funding, and staff. Langford lacked the means to improve the land or properly protect the park, and without formal policy or regulations, he had few legal methods to enforce such protection. This left Yellowstone vulnerable to poachers, vandals, and others seeking to raid its resources. He addressed the practical problems park administrators faced in the 1872 Report to the Secretary of the Interior and correctly predicted that Yellowstone would become a major international attraction deserving the continuing stewardship of the government. In 1874, both Langford and Delano advocated the creation of a federal agency to protect the vast park, but Congress refused. In 1875, Colonel William Ludlow, who had previously explored areas of Montana under the command of George Armstrong Custer, was assigned to organize and lead an expedition to Montana and the newly established Yellowstone Park. Observations about the lawlessness and exploitation of park resources were included in Ludlow's Report of a Reconnaissance to the Yellowstone National Park. The report included letters and attachments by other expedition members, including naturalist and mineralogist George Bird Grinnell.

Grinnell documented the poaching of buffalo, deer, elk, and antelope for hides. "It is estimated that during the winter of 1874–1875, not less than 3,000 buffalo and mule deer suffer even more severely than the elk, and the antelope nearly as much."

As a result, Langford was forced to step down in 1877. Having traveled through Yellowstone and witnessed land management problems first hand, Philetus Norris volunteered for the position following Langford's exit. Congress finally saw fit to implement a salary for the position, as well as to provide a minimal funding to operate the park. Norris used these funds to expand access to the park, building numerous crude roads and facilities.

In 1880, Harry Yount was appointed as a gamekeeper to control poaching and vandalism in the park. Yount had previously spent decades exploring the mountain country of present-day Wyoming, including the Grand Tetons, after joining F V. Hayden's

Geological Survey in 1873. Yount is the first national park ranger, and Yount's Peak, at the head of the Yellowstone River, was named in his honor. However, these measures still proved to be insufficient in protecting the park, as neither Norris, nor the three superintendents who followed, were given sufficient manpower or resources.

Fort Yellowstone.

The Northern Pacific Railroad built a train station in Livingston, Montana, connecting to the northern entrance in the early 1880s, which helped to increase visitation from 300 in 1872 to 5,000 in 1883. Visitors in these early years faced poor roads and limited services, and most access into the park was on horse or via stagecoach. By 1908 visitation increased enough to attract a Union Pacific Railroad connection to West Yellowstone, though rail visitation fell off considerably by World War II and ceased around the 1960s. Much of the railroad line was converted to nature trails, among them the Yellowstone Branch Line Trail.

Thomas Moran painted Tower Creek.

During the 1870s and 1880s, Native American tribes were effectively excluded from the national park. Under a half-dozen tribes had made seasonal use of the Yellowstone area, but the only year-round residents were small bands of Eastern Shoshone known as "Sheepeaters". They left the area under the assurances of a treaty negotiated in 1868, under which the Sheepeaters ceded their lands but retained the right to hunt in Yellowstone. The United States never ratified the treaty and refused to recognize the claims of the Sheepeaters or any other tribe that had used Yellowstone.

The Nez Perce band associated with Chief Joseph, numbering about 750 people, passed through Yellowstone National Park in thirteen days during late August 1877. They were being pursued by the U.S. Army and entered the national park about two weeks after the Battle of the Big Hole. Some of the Nez Perce were friendly to the tourists and other people they encountered in the park; some were not. Nine park visitors were briefly taken captive. Despite Joseph and other chiefs ordering that no one should be harmed, at least two people were killed and several wounded. One of the areas where encounters occurred was in Lower Geyser Basin and east along a branch of the Firehole River to Mary Mountain and beyond. That stream is still known as Nez Perce Creek. A group of Bannocks entered the park in 1878, alarming park Superintendent Philetus Norris. In the aftermath of the Sheepeater Indian War of 1879, Norris built a fort to prevent Native Americans from entering the national park.

Ongoing poaching and destruction of natural resources continued unabated until the U.S. Army arrived at Mammoth Hot Springs in 1886 and built Camp Sheridan. Over the next 22 years the army constructed permanent structures, and Camp Sheridan was renamed Fort Yellowstone. The Boone and Crockett Club, acting through the personality of George G. Vest, Arnold Hague, William Hallett Phillips, W. A. Wadsworth, Archibald Rogers, Theodore Roosevelt, and George Bird Grinnell were successful in carrying through the Park Protection Act, which saved the park. The Lacey Act of 1900 provided legal support for the officials prosecuting poachers. With the funding and manpower necessary to keep a diligent watch, the army developed their own policies and regulations that permitted public access while protecting park wildlife and natural resources. When the National Park Service was created in 1916, many of the management principles developed by the army were adopted by the new agency. The army turned control over to the National Park Service on October 31, 1918.

In 1898, the naturalist John Muir described the park as follows: "However orderly your excursions or aimless, again and again amid the calmest, stillest scenery you will be brought to a standstill hushed and awe-stricken before phenomena wholly new to you. Boiling springs and huge deep pools of purest green and azure water, thousands of them, are plashing and heaving in these high, cool mountains as if a fierce furnace fire were burning beneath each one of them; and a hundred geysers, white torrents of boiling water and steam, like inverted waterfalls, are ever and anon rushing up out of the hot, black underworld."

Superintendent Horace M. Albright and black bears.

By 1915, 1,000 automobiles per year were entering the park, resulting in conflicts with horses and horse-drawn transportation. Horse travel on roads was eventually prohibited.

The Civilian Conservation Corps (CCC), a New Deal relief agency for young men, played a major role between 1933 and 1942 in developing Yellowstone facilities. CCC projects included reforestation, campground development of many of the park's trails and campgrounds, trail construction, fire hazard reduction, and fire-fighting work. The CCC built the majority of the early visitor centers, campgrounds and the current system of park roads.

During World War II, tourist travel fell sharply, staffing was cut, and many facilities fell into disrepair. By the 1950s, visitation increased tremendously in Yellowstone and other national parks. To accommodate the increased visitation, park officials implemented Mission 66, an effort to modernize and expand park service facilities. Planned to be completed by 1966, in honor of the 50th anniversary of the founding of the National Park Service, Mission 66 construction diverged from the traditional log cabin style with design features of a modern style. During the late 1980s, most construction styles in Yellowstone reverted to the more traditional designs. After the enormous forest fires of 1988 damaged much of Grant Village, structures there were rebuilt in the traditional style. The visitor center at Canyon Village, which opened in 2006, incorporates a more traditional design as well.

The Roosevelt Arch.

The 1959 Hebgen Lake earthquake just west of Yellowstone at Hebgen Lake damaged roads and some structures in the park. In the northwest section of the park, new geysers were found, and many existing hot springs became turbid. It was the most powerful earthquake to hit the region in recorded history.

In 1963, after several years of public controversy regarding the forced reduction of the elk population in Yellowstone, United States Secretary of the Interior Stewart Udall appointed an advisory board to collect scientific data to inform future wildlife management of the national parks. In a paper known as the Leopold Report, the committee observed that culling programs at other national parks had been ineffective, and recommended management of Yellowstone's elk population.

The wildfires during the summer of 1988 were the largest in the history of the park. Approximately 793,880 acres (3,210 km^2; 1,240 sq mi) or 36% of the parkland was impacted by the fires, leading to a systematic re-evaluation of fire management policies.

The fire season of 1988 was considered normal until a combination of drought and heat by mid-July contributed to an extreme fire danger. On "Black Saturday", August 20, 1988, strong winds expanded the fires rapidly, and more than 150,000 acres (610 km^2; 230 sq mi) burned.

Pictorial map.

The expansive cultural history of the park has been documented by the 1,000 archeological sites that have been discovered. The park has 1,106 historic structures and features, and of these Obsidian Cliff and five buildings have been designated National Historic Landmarks. Yellowstone was designated an International Biosphere Reserve and a UN World Heritage Site. The park was placed on the List of World Heritage in Danger from 1995 to 2003 due to the effects of tourism, infection of wildlife, and issues with invasive species. In 2010, Yellowstone National Park was honored with its own quarter under the America the Beautiful Quarters Program.

Justin Ferrell explores three moral sensibilities that motivated activists in dealing with Yellowstone. First came the utilitarian vision of maximum exploitation of natural resources, characteristic of developers in the late 19th century. Second was the spiritual vision of nature inspired by the Romanticism and the transcendentalists in the mid-19th century. The twentieth century saw the biocentric moral vision that focuses on the health of the ecosystem as theorized by Aldo Leopold, which led to the expansion of federally protected areas and to the surrounding ecosystems.

Heritage and Research Center

The Heritage and Research Center is located at Gardiner, Montana, near the north entrance to the park. The center is home to the Yellowstone National Park's museum collection, archives, research library, historian, archeology lab, and herbarium. The Yellowstone National Park Archives maintain collections of historical records of Yellowstone and the National Park Service. The collection includes the administrative records of Yellowstone, as well as resource management records, records from major projects, and donated manuscripts and personal papers. The archives are affiliated with the National Archives and Records Administration.

Geography

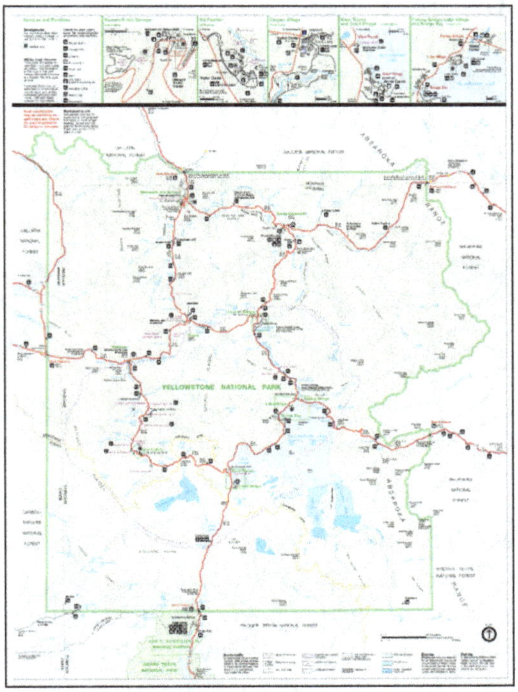

Official park map.

Approximately 96 percent of the land area of Yellowstone National Park is located within the state of Wyoming. Another three percent is within Montana, with the remaining one percent in Idaho. The park is 63 miles (101 km) north to south, and 54 miles (87 km) west to east by air. Yellowstone is 2,219,789 acres (8,983 km²; 3,468 sq mi) in area, larger than the states of Rhode Island or Delaware. Rivers and lakes cover five percent of the land area, with the largest water body being Yellowstone Lake at 87,040 acres (352 km²; 136 sq mi). Yellowstone Lake is up to 400 feet (120 m) deep and has 110 miles (180 km) of shoreline. At an elevation of 7,733 feet (2,357 m) above sea level, Yellowstone Lake is the largest high elevation lake in North America. Forests comprise 80 percent of the land area of the park; most of the rest is grassland.

The Continental Divide of North America runs diagonally through the southwestern part of the park. The divide is a topographic feature that separates Pacific Ocean and Atlantic Ocean water drainages. About one third of the park lies on the west side of the divide. The origins of the Yellowstone and Snake Rivers are near each other but on opposite sides of the divide. As a result, the waters of the Snake River flow to the Pacific Ocean, while those of the Yellowstone find their way to the Atlantic Ocean via the Gulf of Mexico.

The park sits on the Yellowstone Plateau, at an average elevation of 8,000 feet (2,400 m) above sea level. The plateau is bounded on nearly all sides by mountain ranges of the Middle Rocky Mountains, which range from 9,000 to 11,000 feet (2,700 to 3,400 m) in elevation. The highest point in the park is atop Eagle Peak (11,358 feet or 3,462 metres) and the lowest is along Reese Creek (5,282 feet or 1,610 metres). Nearby mountain ranges include the Gallatin Range to the northwest, the Beartooth Mountains in the north, the Absaroka Range to the east, the Teton Range to the south, and the Madison Range to the west. The most prominent summit on the Yellowstone Plateau is Mount Washburn at 10,243 feet (3,122 m).

A 3D computer-generated aerial image.

Yellowstone National Park has one of the world's largest petrified forests, trees which were long ago buried by ash and soil and transformed from wood to mineral materials. This ash and other volcanic debris are believed to have come from the park area itself as the central part of Yellowstone is the massive caldera of a supervolcano. The park contains 290 waterfalls of at least 15 feet (4.6 m), the highest being the Lower Falls of the Yellowstone River at 308 feet (94 m).

Three deep canyons are located in the park, cut through the volcanic tuff of the Yellowstone Plateau by rivers over the last 640,000 years. The Lewis River flows through Lewis Canyon in the south, and the Yellowstone River has carved two colorful canyons, the Grand Canyon of the Yellowstone and the Black Canyon of the Yellowstone in its journey north.

Geology

Volcanism

Columnar basalt near Tower Fall; large floods of basalt and other lava types preceded mega-eruptions of superheated ash and pumice.

Yellowstone is at the northeastern end of the Snake River Plain, a great U-shaped arc through the mountains that extends from Boise, Idaho some 400 miles (640 km) to the west.

The volcanism of Yellowstone is believed to be linked to the somewhat older volcanism of Snake River plain. Yellowstone is thus the active part of a hotspot that has moved northeast over time. The origin of this hotspot volcanism is disputed. One theory holds that a mantle plume has caused the Yellowstone hotspot to migrate northeast, while another theory explains migrating hotspot volcanism as the result of the fragmentation and dynamics of the subducted Farallon Plate in Earth's interior.

The Yellowstone Caldera is the largest volcanic system in North America. It has been termed a "supervolcano" because the caldera was formed by exceptionally large explosive eruptions. The magma chamber that lies under Yellowstone is estimated to be a single connected chamber, about 37 miles (60 km) long, 18 miles (29 km) wide, and 3 to 7 miles (5 to 12 km) deep. The current caldera was created by a cataclysmic eruption that occurred 640,000 years ago, which released more than 240 cubic miles (1,000 km^3) of ash, rock and pyroclastic materials. This eruption was more than 1,000 times larger than the 1980 eruption of Mount St. Helens. It produced a caldera nearly 5/8 of a mile (1 km) deep and 45 by 28 miles (72 by 45 km) in area and deposited the Lava Creek Tuff, a welded tuff geologic formation. The most violent known eruption, which occurred 2.1 million years ago, ejected 588 cubic miles (2,450 km^3) of volcanic material and created the rock formation known as the Huckleberry Ridge Tuff and created the Island Park Caldera. A smaller eruption ejected 67 cubic miles (280 km^3) of material 1.3 million years ago, forming the Henry's Fork Caldera and depositing the Mesa Falls Tuff.

Each of the three climactic eruptions released vast amounts of ash that blanketed much of central North America, falling many hundreds of miles away. The amount of ash and gases released into the atmosphere probably caused significant impacts to world weather patterns and led to the extinction of some species, primarily in North America.

Boardwalks allow visitors to safely approach the thermal features, such as Grand Prismatic Spring.

A subsequent caldera-forming eruption occurred about 160,000 years ago. It formed the relatively small caldera that contains the West Thumb of Yellowstone Lake. Since the last supereruption, a series of smaller eruptive cycles between 640,000 and 70,000 years ago, has nearly filled in the Yellowstone Caldera with 80 different eruptions of rhyolitic lavas such as those that can be seen at Obsidian Cliffs and basaltic lavas which can be viewed at Sheepeater Cliff. Lava strata are most easily seen at the Grand Canyon of the Yellowstone, where the Yellowstone River continues to carve into the ancient lava flows. The canyon is a classic V-shaped valley, indicative of river-type erosion rather than erosion caused by glaciation.

Each eruption is part of an eruptive cycle that climaxes with the partial collapse of the roof of the volcano's partially emptied magma chamber. This creates a collapsed depression, called a caldera, and releases vast amounts of volcanic material, usually through fissures that ring the caldera. The time between the last three cataclysmic eruptions in the Yellowstone area has ranged from 600,000 to 800,000 years, but the small number of such climactic eruptions cannot be used to make an accurate prediction for future volcanic events.

Geysers and the Hydrothermal System

Old Faithful erupts approximately every 90 minutes.

The most famous geyser in the park, and perhaps the world, is Old Faithful geyser, located in Upper Geyser Basin. Castle Geyser, Lion Geyser and Beehive Geyser are in the same basin. The park contains the largest active geyser in the world—Steamboat Geyser in the Norris Geyser Basin. A study that was completed in 2011 found that at least 1283 geysers have erupted in Yellowstone. Of these, an average of 465 are active in a given year. Yellowstone contains at least 10,000 thermal features altogether. Half of the world's geysers and hydrothermal features are concentrated in Yellowstone.

The U.S. Geological Survey, Yellowstone National Park, and the University of Utah created the Yellowstone Volcano Observatory (YVO), a partnership for long-term monitoring of the geological processes of the Yellowstone Plateau volcanic field, for disseminating information concerning the potential hazards of this geologically active region.

Steamboat Geyser is the world's largest active geyser.

In 2003, changes at the Norris Geyser Basin resulted in the temporary closure of some trails in the basin. New fumaroles were observed, and several geysers showed enhanced activity and increasing water temperatures. Several geysers became so hot that they were transformed into purely steaming features; the water had become superheated and they could no longer erupt normally. This coincided with the release of reports of a multiple year United States Geological Survey research project which mapped the bottom of Yellowstone Lake and identified a structural dome that had uplifted at some time in the past. Research indicated that these uplifts posed no immediate threat of a volcanic eruption, since they may have developed long ago, and there had been no temperature increase found near the uplifts. On March 10, 2004, a biologist discovered 5 dead bison which apparently had inhaled toxic geothermal gases trapped in the Norris Geyser Basin by a seasonal atmospheric inversion. This was closely followed by an upsurge of earthquake activity in April 2004. In 2006, it was reported that the Mallard Lake Dome and the Sour Creek Dome—areas that have long been known to show significant changes in their ground movement—had risen at a rate of 1.5 to 2.4 inches (3.8 to 6.1 cm) per year from mid–2004 through 2006. As of late 2007, the uplift has continued at a reduced rate. These events inspired a great deal of media attention and speculation about the geologic future of the region. Experts responded to the conjecture by informing the public that there was no increased risk of a volcanic eruption in the near future. However, these changes demonstrate the dynamic nature of the Yellowstone hydrothermal system.

Earthquakes

Yellowstone experiences thousands of small earthquakes every year, virtually all of which are undetectable to people. There have been six earthquakes with at least magnitude 6 or greater in historical times, including the 7.2magnitude Hebgen Lake earthquake which occurred just outside the northwest boundary of the park in 1959. This quake triggered a huge landslide, which caused a partial dam collapse on Hebgen Lake; immediately downstream, the sediment from the landslide dammed the river and created a new lake, known as Earthquake Lake. Twenty-eight people were killed, and property damage was extensive in the immediate region. The earthquake caused some geysers in the northwestern section of the park to erupt, large cracks in the ground formed and emitted steam, and some hot springs that normally have clear water turned muddy. A 6.1magnitude earthquake struck inside the park on June 30, 1975, but damage was minimal.

Infrastructure damage at Hebgen Lake due to the 7.2 magnitude earthquake.

For three months in 1985, 3,000 minor earthquakes were detected in the northwestern section of the park, during what has been referred to as an earthquake swarm, and has been attributed to minor subsidence of the Yellowstone caldera. Beginning on April 30, 2007, 16 small earthquakes with magnitudes up to 2.7 occurred in the Yellowstone Caldera for several days. These swarms of earthquakes are common, and there have been 70 such swarms between 1983 and 2008. In December 2008, over 250 earthquakes were measured over a four-day span under Yellowstone Lake, the largest measuring a magnitude of 3.9. In January 2010, more than 250 earthquakes were detected over a two-day period. Seismic activity in Yellowstone National Park continues and is reported hourly by the Earthquake Hazards Program of the U.S. Geological Survey.

On March 30, 2014, a magnitude 4.8 earthquake struck almost the very middle of Yellowstone near the Norris Basin at 6:34 am; reports indicated no damage. This was the largest earthquake to hit the park since February 22, 1980.

Biology and Ecology

Meadow in Yellowstone National Park.

Yellowstone National Park is the centerpiece of the 20 million acres (80,940 km^2; 31,250 sq mi) Greater Yellowstone Ecosystem, a region that includes Grand Teton National Park, adjacent National Forests and expansive wilderness areas in those forests. The ecosystem is the largest remaining continuous stretch of mostly undeveloped pristine land in the contiguous United States, considered the world's largest intact ecosystem in the northern temperate zone. With the successful wolf reintroduction program, which began in the 1990s, virtually all the original faunal species known to inhabit the region when white explorers first entered the area can still be found there.

Flora

Over 1,700 species of trees and other vascular plants are native to the park. Another 170 species are considered to be exotic species and are non-native. Of the eight conifer tree species documented, Lodgepole Pine forests cover 80% of the total forested areas. Other conifers, such as Subalpine Fir, Engelmann Spruce, Rocky Mountain Douglas-fir and Whitebark Pine, are found in scattered groves throughout the park. As of 2007, the whitebark pine is threatened by a fungus known as white pine blister rust; however, this is mostly confined to forests well to the north and west. In Yellowstone, about seven percent of the whitebark pine species have been impacted with the fungus, compared to nearly complete infestations in northwestern Montana. Quaking Aspen and willows are the most common species of deciduous trees. The aspen forests have declined significantly since the early 20th century, but scientists at Oregon State University attribute recent recovery of the aspen to the reintroduction of wolves which has changed the grazing habits of local elk.

Yellowstone sand verbena are endemic to Yellowstone's lakeshores.

There are dozens of species of flowering plants that have been identified, most of which bloom between the months of May and September. The Yellowstone sand verbena is a rare flowering plant found only in Yellowstone. It is closely related to species usually found in much warmer climates, making the sand verbena an enigma. The estimated 8,000 examples of this rare flowering plant all make their home in the sandy soils on the shores of Yellowstone Lake, well above the waterline.

In Yellowstone's hot waters, bacteria form mats of bizarre shapes consisting of trillions of individuals. These bacteria are some of the most primitive life forms on earth. Flies and other arthropods live on the mats, even in the middle of the bitterly cold winters. Initially, scientists thought that microbes there gained sustenance only from sulfur. In 2005 researchers from the University of Colorado at Boulder discovered that the sustenance for at least some of the diverse hyperthermophilic species is molecular hydrogen.

Thermus aquaticus is a bacterium found in the Yellowstone hot springs that produces an important enzyme (Taq polymerase) that is easily replicated in the lab and is useful in replicating DNA as part of the polymerase chain reaction (PCR) process. The retrieval of these bacteria can be achieved with no impact to the ecosystem. Other bacteria in the Yellowstone hot springs may also prove useful to scientists who are searching for cures for various diseases. In 2016, researchers from Uppsala University reported the discovery of a class of thermophiles, Hadesarchaea, in Yellowstone's Culex Basin. These organisms are capable of converting carbon monoxide and water to carbon dioxide and oxygen.

Non-native plants sometimes threaten native species by using up nutrient resources. Though exotic species are most commonly found in areas with the greatest human visitation, such as near roads and at major tourist areas, they have also spread into the backcountry. Generally, most exotic species are controlled by pulling the plants out of the soil or by spraying, both of which are time consuming and expensive.

Fauna

Yellowstone is widely considered to be the finest megafauna wildlife habitat in the lower 48 states. There are almost 60 species of mammals in the park, including the Rocky Mountain wolf, coyote, the threatened Canadian lynx, and grizzly bears. Other large mammals include the bison (often referred to as buffalo), black bear, elk, moose, mule deer, white-tailed deer, mountain goat, pronghorn, bighorn sheep, and cougar.

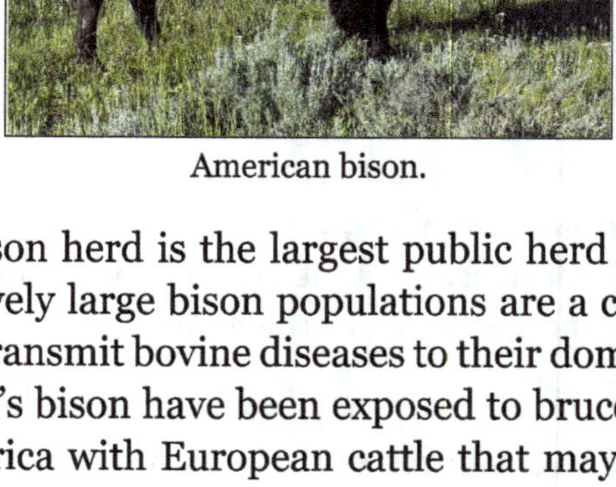

American bison.

The Yellowstone Park bison herd is the largest public herd of American bison in the United States. The relatively large bison populations are a concern for ranchers, who fear that the species can transmit bovine diseases to their domesticated cousins. In fact, about half of Yellowstone's bison have been exposed to brucellosis, a bacterial disease that came to North America with European cattle that may cause cattle to miscarry.

The disease has little effect on park bison, and no reported case of transmission from wild bison to domestic livestock has been filed. However, the Animal and Plant Health Inspection Service (APHIS) has stated that bison are the "likely source" of the spread of the disease in cattle in Wyoming and North Dakota. Elk also carry the disease and are believed to have transmitted the infection to horses and cattle. Bison once numbered between 30 and 60 million individuals throughout North America, and Yellowstone remains one of their last strongholds. Their populations had increased from less than 50 in the park in 1902 to 4,000 by 2003. The Yellowstone Park bison herd reached a peak in 2005 with 4,900 animals. Despite a summer estimated population of 4,700 in 2007, the number dropped to 3,000 in 2008 after a harsh winter and controversial brucellosis management sending hundreds to slaughter. The Yellowstone Park bison herd is believed to be one of only four free roaming and genetically pure herds on public lands in North America. The other three herds are the Henry Mountains bison herd of Utah, at Wind Cave National Park in South Dakota and in Elk Island National Park in Alberta.

Elk mother nursing her calf.

To combat the perceived threat of brucellosis transmission to cattle, national park personnel regularly harass bison herds back into the park when they venture outside of the area's borders. During the winter of 1996–97, the bison herd was so large that 1,079 bison that had exited the park were shot or sent to slaughter. Animal rights activists argue that this is a cruel practice and that the possibility for disease transmission is not as great as some ranchers maintain. Ecologists point out that the bison are merely traveling to seasonal grazing areas that lie within the Greater Yellowstone Ecosystem that have been converted to cattle grazing, some of which are within National Forests and are leased to private ranchers. APHIS has stated that with vaccinations and other means, brucellosis can be eliminated from the bison and elk herds throughout Yellowstone.

A reintroduced wolf in Yellowstone National Park.

Starting in 1914, in an effort to protect elk populations, the U.S. Congress appropriated funds to be used for the purposes of "destroying wolves, prairie dogs, and other animals injurious to agriculture and animal husbandry" on public lands. Park Service hunters carried out these orders, and by 1926 they had killed 136 wolves, and wolves were virtually eliminated from Yellowstone. Further exterminations continued until the National Park Service ended the practice in 1935. With the passing of the Endangered Species Act in 1973, the wolf was one of the first mammal species listed. After the wolves were extirpated from Yellowstone, the coyote then became the park's top canine predator. However, the coyote is not able to bring down large animals, and the result of this lack of a top predator on these populations was a marked increase in lame and sick megafauna.

By the 1990s, the Federal government had reversed its views on wolves. In a controversial decision by the U.S. Fish and Wildlife Service (which oversees threatened and endangered species), northwestern wolves imported from Canada were reintroduced into the park. Reintroduction efforts have been successful with populations remaining relatively stable. A survey conducted in 2005 reported that there were 13 wolf packs, totaling 118 individuals in Yellowstone and 326 in the entire ecosystem. These park figures were lower than those reported in 2004 but may be attributable to wolf migration to other nearby areas as suggested by the substantial increase in the Montana population during that interval. Almost all the wolves documented were descended from the 66 wolves reintroduced in 1995–96. The recovery of populations throughout the states of Wyoming, Montana and Idaho has been so successful that on February 27, 2008, the U.S. Fish and Wildlife Service removed the Northern Rocky Mountain wolf population from the endangered species list.

Black bear and cub near Tower Fall.

Black bears are common in the park and were a park symbol due to visitor interaction with the bears starting in 1910. Feeding and close contact with bears has not been permitted since the 1960s to reduce their desire for human foods. Yellowstone is one of the few places in the United States where black bears can be seen coexisting with grizzly bears. Black bear observations occur most often in the park's northern ranges and in the Bechler area which is in the park's southwestern corner.

As of 2017, an estimated 700 grizzly bears were living in the Greater Yellowstone Ecosystem, with about 150 grizzlies living wholly or partially within Yellowstone National

Park. The grizzly was initially listed as a threatened species in the contiguous United States on July 28, 1975, by the Fish and Wildlife Service. The grizzly bear was taken off the endangered species list in 2007. Opponents of delisting the grizzly expressed concerns that states might once again allow hunting and that better conservation measures were needed to ensure a sustainable population. A federal district judge overturned the delisting ruling in 2009, reinstating the grizzly; however, the grizzly was once again removed from the list in 2017. In September 2018, a US district judge ruled that the grizzly's protections must be restored in full, arguing the Fish and Wildlife Service was mistaken in removing the bear from the threatened status list. Regardless of the rulings, hunting is prohibited within Yellowstone National Park. Hunters who legally hunt animals outside park boundaries may transport the carcass through the park with a permit.

Elk in Hayden Valley.

Population figures for elk are in excess of 30,000—the largest population of any large mammal species in Yellowstone. The northern herd has decreased enormously since the mid1990s; this has been attributed to wolf predation and causal effects such as elk using more forested regions to evade predation, consequently making it harder for researchers to accurately count them. The northern herd migrates west into southwestern Montana in the winter. The southern herd migrates southward, and the majority of these elk winter on the National Elk Refuge, immediately southeast of Grand Teton National Park. The southern herd migration is the largest mammalian migration remaining in the U.S. outside of Alaska.

Pronghorn are commonly found on the grasslands in the park.

In 2003 the tracks of one female lynx and her cub were spotted and followed for over 2 miles (3.2 km). Fecal material and other evidence obtained were tested and confirmed to be those of a lynx. No visual confirmation was made, however. Lynx have not been seen in Yellowstone since 1998, though DNA taken from hair samples obtained in 2001 confirmed that lynx were at least transient to the park. Other less commonly seen mammals include the mountain lion and wolverine. The mountain lion has an estimated population of only 25 individuals parkwide. The wolverine is another rare park mammal, and accurate population figures for this species are not known. These uncommon and rare mammals provide insight into the health of protected lands such as Yellowstone and help managers make determinations as to how best to preserve habitats.

Eighteen species of fish live in Yellowstone, including the core range of the Yellowstone cutthroat trout—a fish highly sought by anglers. The Yellowstone cutthroat trout has faced several threats since the 1980s, including the suspected illegal introduction into Yellowstone Lake of lake trout, an invasive species which consume the smaller cutthroat trout. Although lake trout were established in Shoshone and Lewis lakes in the Snake River drainage from U.S. Government stocking operations in 1890, it was never officially introduced into the Yellowstone River drainage. The cutthroat trout has also faced an ongoing drought, as well as the accidental introduction of a parasite—whirling disease—which causes a terminal nervous system disease in younger fish. Since 2001, all native sport fish species caught in Yellowstone waterways are subject to a catch and release law. Yellowstone is also home to six species of reptiles, such as the painted turtle, boa prairie rattlesnake, and four species of amphibians, including the boreal chorus frog.

Three hundred eleven species of birds have been reported, almost half of which nest in Yellowstone. In 1999, twenty-six pairs of nesting bald eagle were documented. Extremely rare sightings of whooping cranes have been recorded, however only three examples of this species are known to live in the Rocky Mountains, out of 385 known worldwide. Other birds, considered to be species of special concern because of their rarity in Yellowstone, include the common loon, harlequin duck, osprey, peregrine falcon and the trumpeter swan.

Forest Fires

As wildfire is a natural part of most ecosystems, plants that are indigenous to Yellowstone have adapted in a variety of ways. Douglas-fir have a thick bark which protects the inner section of the tree from most fires. Lodgepole Pines—the most common tree species in the park—generally have cones that are only opened by the heat of fire. Their seeds are held in place by a tough resin, and fire assists in melting the resin, allowing the seeds to disperse. Fire clears out dead and downed wood, providing fewer obstacles for lodgepole pines to flourish. Subalpine Fir, Engelmann Spruce, Whitebark Pine, and other species tend to grow in colder and moister areas, where fire is less likely to occur. Aspen trees sprout new growth from their roots, and even if a severe fire kills the tree above ground, the roots often survive unharmed because they are insulated from the heat by soil. The National Park Service estimates that in natural conditions, grasslands in Yellowstone burned an average of every 20 to 25 years, while forests in the park would experience fire about every 300 years.

Fire in Yellowstone National Park.

About thirty-five natural forest fires are ignited each year by lightning, while another six to ten are started by people—in most cases by accident. Yellowstone National Park has three fire lookout towers, each staffed by trained fire fighters. The easiest one to reach is atop Mount Washburn, which has interpretive exhibits and an observation deck open to the public. The park also monitors fire from the air and relies on visitor reports of smoke and flames. Fire towers are staffed almost continuously from late June to mid-September—the primary fire season. Fires burn with the greatest intensity in the late afternoon and evening. Few fires burn more than 100 acres (40 ha), and the vast majority of fires reach only a little over an acre (0.5 ha) before they burn themselves out. Fire management focuses on monitoring dead and down wood quantities, soil and tree moisture, and the weather, to determine those areas most vulnerable to fire should one ignite. Current policy is to suppress all human caused fires and to evaluate natural fires, examining the benefit or detriment they may pose on the ecosystem. If a fire is considered to be an immediate threat to people and structures, or will burn out of control, then fire suppression is performed.

Wildfire in Yellowstone National Park produces a pyrocumulus cloud.

In an effort to minimize the chances of out of control fires and threats to people and structures, park employees do more than just monitor the potential for fire. Controlled burns are prescribed fires which are deliberately started to remove dead timber under conditions which allow fire fighters an opportunity to carefully control where and how much wood is consumed. Natural fires are sometimes considered prescribed fires if they are left to burn. In Yellowstone, unlike some other parks, there have been very few fires deliberately started by employees as prescribed burns. However, over the last 30 years, over 300 natural fires have been allowed to burn naturally. In addition, fire fighters remove dead and down wood and other hazards from areas where they will be a potential fire threat to lives and property, reducing the chances of fire danger in these areas. Fire monitors also regulate fire through educational services to the public and have been known to temporarily ban campfires from campgrounds during periods of high fire danger. The common notion in early United States land management policies was that all forest fires were bad. Fire was seen as a purely destructive force and there was little understanding that it was an integral part of the ecosystem. Consequently, until the 1970s, when a better understanding of wildfire was developed, all fires were suppressed. This led to an increase in dead and dying forests, which would later provide the fuel load for fires that would be much harder, and in some cases, impossible to control. Fire Management Plans were implemented, detailing that natural fires should be allowed to burn if they posed no immediate threat to lives and property.

A crown fire approaches the Old Faithful complex.

1988 started with a wet spring season although by summer, drought began moving in throughout the northern Rockies, creating the driest year on record to that point. Grasses and plants which grew well in the early summer from the abundant spring moisture produced plenty of grass, which soon turned to dry tinder. The National Park Service began firefighting efforts to keep the fires under control, but the extreme drought made suppression difficult. Between July 15 and 21, 1988, fires quickly spread from 8,500 acres (3,400 ha; 13.3 sq mi) throughout the entire Yellowstone region, which included areas outside the park, to 99,000 acres (40,000 ha; 155 sq mi) on the park land alone. By the end of the month, the fires were out of control. Large fires burned together, and on August 20, 1988, the single worst day of the fires, more than 150,000 acres

(61,000 ha; 230 sq mi) were consumed. Seven large fires were responsible for 95% of the 793,000 acres (321,000 ha; 1,239 sq mi) that were burned over the next couple of months. A total of 25,000 firefighters and U.S. military forces participated in the suppression efforts, at a cost of 120 million dollars. By the time winter brought snow that helped extinguish the last flames, the fires had destroyed 67 structures and caused several million dollars in damage. Though no civilian lives were lost, two personnel associated with the firefighting efforts were killed.

Contrary to media reports and speculation at the time, the fires killed very few park animals—surveys indicated that only about 345 elk (of an estimated 40,000–50,000), 36 deer, 12 moose, 6 black bears, and 9 bison had perished. Changes in fire management policies were implemented by land management agencies throughout the United States, based on knowledge gained from the 1988 fires and the evaluation of scientists and experts from various fields. By 1992, Yellowstone had adopted a new fire management plan which observed stricter guidelines for the management of natural fires.

Climate

Winter scene in Yellowstone.

Yellowstone climate is greatly influenced by altitude, with lower elevations generally found to be warmer year-round. The record high temperature was 99 °F (37 °C) in 2002, while the coldest temperature recorded is −66 °F (−54 °C) in 1933. During the summer months of June to early September, daytime highs are normally in the 70 to 80 °F (21 to 27 °C) range, while night time lows can go to below freezing (0 °C) especially at higher altitudes. Summer afternoons are frequently accompanied by thunderstorms. Spring and fall temperatures range between 30 and 60 °F (−1 and 16 °C) with nights in the teens to single digits (−5 to −20 °C). Winter in Yellowstone is accompanied by high temperatures usually between zero and 20 °F (−20 to −5 °C) and nighttime temperatures below 0 °F (−18 °C) for most of the winter.

Precipitation in Yellowstone is highly variable and ranges from 15 inches (380 mm) annually near Mammoth Hot Springs, to 80 inches (2,000 mm) in the southwestern sections of the park. The precipitation of Yellowstone is greatly influenced by the moisture channel formed by the Snake River Plain to the west that was, in turn, formed by Yellowstone itself. Snow is possible in any month of the year, but most common between November and April, with averages of 150 inches (3,800 mm) annually around Yellowstone Lake, to twice that amount at higher elevations.

Tornadoes in Yellowstone are rare; however, the most powerful tornado recorded in Wyoming touched down in the Teton Wilderness of Bridger-Teton National Forest and hit Yellowstone National Park. Called the Teton–Yellowstone tornado, it was classified as an F4, with wind speeds estimated at between 207 and 260 miles per hour (333 and 418 km/h). The tornado left a path of destruction 1 to 2 miles (1.6 to 3.2 km) wide, and 24 miles (39 km) long, and leveled 15,000 acres (6,100 ha; 23 sq mi) of mature pine forest.

The climate at Yellowstone Lake is classified as subarctic (Dfc), according to Köppen-Geiger climate classification, while at the park headquarters the classification is humid continental (Dfb).

Olympic National Park

Olympic National Park is an American national park located in the State of Washington, on the Olympic Peninsula. The park has four regions: the Pacific coastline, alpine areas, the west side temperate rainforest and the forests of the drier east side. Within the park there are three distinct ecosystems which are subalpine forest and wildflower meadow, temperate forest, and the rugged Pacific coast.

President Theodore Roosevelt originally designated Mount Olympus National Monument on 2 March 1909. The monument was redesignated as a national park by Congress and President Franklin Roosevelt on June 29, 1938. In 1976, Olympic National Park was designated by UNESCO as an International Biosphere Reserve, and in 1981 as a World Heritage Site. In 1988, Congress designated 95 percent of the park as the Olympic Wilderness.

Coastline

Tide pools form at low tide.

The coastal portion of the park is a rugged, sandy beach along with a strip of adjacent forest. It is 60 miles (97 km) long but just a few miles wide, with native communities at the mouths of two rivers. The Hoh River has the Hoh people and at the town of La Push at the mouth of the Quileute River live the Quileute.

The beach has unbroken stretches of wilderness ranging from 10 to 20 miles (16 to 32 km). While some beaches are primarily sand, others are covered with heavy rock and very large boulders. Bushy overgrowth, slippery footing, tides and misty rain forest weather all hinder foot travel. The coastal strip is more readily accessible than the interior of the Olympics; due to the difficult terrain, very few backpackers venture beyond casual day-hiking distances.

The most popular piece of the coastal strip is the 9-mile (14 km) Ozette Loop. The Park Service runs a registration and reservation program to control usage levels of this area. From the trailhead at Ozette Lake, a 3-mile (4.8 km) leg of the trail is a boardwalk-enhanced path through near primal coastal cedar swamp. Arriving at the ocean, it is a 3-mile walk supplemented by headland trails for high tides. This area has traditionally been favored by the Makah from Neah Bay. The third 3-mile leg is enabled by a boardwalk which has enhanced the loop's visitor numbers.

There are thick groves of trees adjacent to the sand, which results in chunks of timber from fallen trees on the beach. The mostly unaltered Hoh River, toward the south end of the park, discharges large amounts of naturally eroded timber and other drift, which moves north, enriching the beaches. The removal of driftwood – logs, dead-heads, tops and root-wads from streams and beaches was a major domestication measure across North America. Even today driftwood deposits form a commanding presence, biologically as well as visually, giving a taste of the original condition of the beach viewable to some extent in early photos. Drift-material often comes from a considerable distance; the Columbia River formerly contributed huge amounts to the Northwest Pacific coasts.

The smaller coastal portion of the park is separated from the larger, inland portion. President Franklin D. Roosevelt originally had supported connecting them with a continuous strip of park land.

A 3D computer generated aerial view.

The park is known for its unique turbidites. It has very exposed turbidities with white calcite veins. Turbidites are rocks or sediments that travel into the ocean as suspended particles in the flow of water, causing a sedimentary layering effect on the ocean floor. Over time the sediments and rock compact and the process repeats as a constant cycle. The park also is known for its tectonic mélanges that have been deemed 'smell rocks' by the locals due to its strong petroleum odor. Mélanges are large individual rocks that are large enough that they are accounted for in map drawings. The Olympic mélanges can be as large as a house.

Glaciated Mountains

Within the center of Olympic National Park rise the Olympic Mountains whose sides and ridgelines are topped with massive, ancient glaciers. The mountains themselves are products of accretionary wedge uplifting related to the Juan De Fuca Plate subduction zone. The geologic composition is a curious mélange of basaltic and oceanic sedimentary rock. The western half of the range is dominated by the peak of Mount Olympus, which rises to 7,965 feet (2,428 m). Mount Olympus receives a large amount of snow, and consequently has the greatest glaciation of any non-volcanic peak in the contiguous United States outside of the North Cascades. It has several glaciers, the largest of which is Hoh Glacier at 3.06 miles (4.93 km) in length. Looking to the east, the range becomes much drier due to the rain shadow of the western mountains. Here, there are numerous high peaks and craggy ridges. The tallest summit of this area is Mount Deception, at 7,788 feet (2,374 m).

Temperate Rainforest

Downed western redcedar.

The western side of the park is mantled by temperate rainforests, including the Hoh Rainforest and Quinault Rainforest, which receive annual precipitation of about 150 inches (380 cm), making this perhaps the wettest area in the continental United States (the island of Kauai in the state of Hawaii gets more rain).

As opposed to tropical rainforests and most other temperate rainforest regions, the rainforests of the Pacific Northwest are dominated by coniferous trees, including Sitka Spruce, Western Hemlock, Coast Douglas-fir and Western redcedar. Mosses coat the bark of these trees and even drip down from their branches in green, moist tendrils.

Valleys on the eastern side of the park also have notable old-growth forest, but the climate is notably drier. Sitka Spruce is absent, trees on average are somewhat smaller, and undergrowth is generally less dense and different in character. Immediately northeast of the park is a rather small rainshadow area where annual precipitation averages about 16 inches.

Ecology

Black-tailed deer.

According to the A. W. Kuchler U.S. Potential natural vegetation Types, Olympic National Park encompasses five classifications: Alpine Meadows and Barren, aka Alpine tundra potential vegetation type with an Alpine Meadow potential vegetation form; a Fir/Hemlock vegetation type with a Pacific Northwest conifer forest vegetation form; a cedar/hemlock/Douglas fir vegetation type with a Pacific Northwest conifer forest vegetation form; Western spruce/fir vegetation type with a Rocky Mountain conifer forest vegetation form; and a spruce/cedar/hemlock vegetation type with a Pacific Northwest conifer forest vegetation form.

Because the park sits on an isolated peninsula, with a high mountain range dividing it from the land to the south, it developed many endemic plant and animal species (like the Olympic Marmot, Piper's bellflower and Flett's violet). The southwestern coastline of the Olympic Peninsula is also the northernmost non-glaciated region on the Pacific coast of North America, with the result that – aided by the distance from peaks to the coast at the Last Glacial Maximum being about twice what it is today – it served as a refuge from which plants colonized glaciated regions to the north.

The park also provides habitat for many species (like the Roosevelt elk) that are native only to the Pacific Northwest coast. As a result, scientists have declared it a biological reserve and study its unique species to better understand how plants and animals evolve. The park is home to sizable populations of black bears and black-tailed deer. The park also has a noteworthy cougar population, numbering about 150. Mountain goats were accidentally introduced into the park into the 1920s and have caused much damage on the native flora. The NPS has activated management plans to control the goats. The park contains an estimated 366,000 acres (572 sq mi; 1,480 km²) of old-growth forests.

Subalpine fir in a meadow on Hurricane Ridge.

Forest fires are infrequent in the rainforests of the park's western side; however, a severe drought after the driest spring in 100 years, coupled with an extremely low snowpack from the preceding winter, resulted in a rare rainforest fire in the summer of 2015.

Ecological zones - Glaciated mountains, subalpine forests and meadows, temperate rainforests, and coastline

Fauna

Shi Shi Beach.

Animals that inhabit this national park are chipmunks, squirrels, skunks, six species of bats, weasels, coyotes, muskrats, fishers, river otters, beavers, red foxes, mountain goats, martens, bobcats, black bears, Canadian lynxes, moles, snowshoe hares, shrews, and cougars. Whales, dolphins, sea lions, seals, and sea otters swim near this park offshore. Birds that fly in this park including raptors are Winter wrens, and Canada jays, Hammond's flycatchers, Wilson's warblers, Blue Grouses, Pine siskins, ravens, spotted owls, Red-breasted nuthatches, Golden-crowned kinglets, Chestnut-backed chickadees,

Swainson's thrushes, Red crossbills, Hermit thrushes, Olive-sided flycatchers, bald eagles, Western tanagers, Northern pygmy owls, Townsend's warblers, Townsend's solitaires, Vaux's swifts, band-tailed pigeons, and evening grosbeaks.

Yosemite National Park

Yosemite National Park is an American national park located in the western Sierra Nevada of Central California, bounded on the southeast by Sierra National Forest and on the northwest by Stanislaus National Forest. The park is managed by the National Park Service and covers an area of 748,436 acres (1,169 sq mi; 3,029 km^2) and sits in four counties: centered in Tuolumne and Mariposa, extending north and east to Mono and south to Madera County. Designated a World Heritage site in 1984, Yosemite is internationally recognized for its granite cliffs, waterfalls, clear streams, giant sequoia groves, lakes, mountains, meadows, glaciers, and biological diversity. Almost 95% of the park is designated wilderness.

On average, about four million people visit Yosemite each year, and most spend the majority of their time in the seven square miles (18 km^2) of Yosemite Valley. The park set a visitation record in 2016, surpassing five million visitors for the first time in its history. Yosemite was central to the development of the national park idea. Galen Clark and others lobbied to protect Yosemite Valley from development, ultimately leading to President Abraham Lincoln's signing the Yosemite Grant in 1864. John Muir led a successful movement to have Congress establish a larger national park by 1890, one which encompassed the valley and its surrounding mountains and forests, paving the way for the National Park System.

Yosemite is one of the largest and least fragmented habitat blocks in the Sierra Nevada, and the park supports a diversity of plants and animals. The park has an elevation range from 2,127 to 13,114 feet (648 to 3,997 m) and contains five major vegetation zones: chaparral and oak woodland, lower montane forest, upper montane forest, subalpine zone, and alpine. Of California's 7,000 plant species, about 50% occur in the Sierra Nevada and more than 20% are within Yosemite. The park contains suitable habitat for more than 160 rare plants, with rare local geologic formations and unique soils characterizing the restricted ranges many of these plants occupy.

The geology of the Yosemite area is characterized by granitic rocks and remnants of older rock. About 10 million years ago, the Sierra Nevada was uplifted and then tilted to form its relatively gentle western slopes and the more dramatic eastern slopes. The uplift increased the steepness of stream and river beds, resulting in the formation of deep, narrow canyons. About one million years ago, snow and ice accumulated, forming glaciers at the higher alpine meadows that moved down the river valleys. Ice thickness in Yosemite Valley may have reached 4,000 feet (1,200 m) during the early glacial episode. The downslope movement of the ice masses cut and sculpted the U-shaped valley that attracts so many visitors to its scenic vistas today.

The name "Yosemite" (meaning "killer" in Miwok) originally referred to the name of a tribe which was driven out of the area (and possibly annihilated) by the Mariposa Battalion. Previously, the area had been called "Ahwahnee" ("big mouth") by indigenous people.

Yosemite Valley has been inhabited for nearly 3,000 years, although humans may have first visited the area as long as 8,000 to 10,000 years ago. The indigenous natives called themselves the Ahwahnechee, meaning "dwellers in Ahwahnee". They are related to the Northern Paiute and Mono tribes. Many tribes visited the area to trade, including nearby Central Sierra Miwoks, who lived along the drainage area of the Tuolumne and Stanislaus Rivers. A major trading route went over Mono Pass and through Bloody Canyon to Mono Lake, just to the east of the Yosemite area. Vegetation and game in the region were similar to that present today; acorns were a staple to their diet, as well as other seeds and plants, salmon and deer.

The California Gold Rush in the mid-19th century dramatically increased travel by European-Americans in the area, causing competition for resources between the regional Paiute and Miwok and the miners and hangers on. In 1851 as part of the Mariposa Wars intended to suppress Native American resistance, United States Army Major Jim Savage led the Mariposa Battalion into the west end of Yosemite Valley. He was pursuing forces of around 200 Ahwahneechee led by Chief Tenaya.

Accounts from this battalion were the first well-documented reports of ethnic Europeans entering Yosemite Valley. Attached to Savage's unit was Dr. Lafayette Bunnell, the company physician, who later wrote about his awestruck impressions of the valley in The Discovery of the Yosemite. Bunnell is credited with naming Yosemite Valley, based on his interviews with Chief Tenaya. Bunnell wrote that Chief Tenaya was the founder of the Ah-wah-nee colony. The Miwok, a neighboring tribe, and most white settlers considered the Ahwahneechee to be especially violent because of their frequent territorial disputes. The Miwok term for the Pai-Ute band was *yohhe'meti,* meaning "they are killers". Correspondence and articles written by members of the battalion helped to popularize the natural wonders of the Yosemite Valley and the surrounding area. Chief Tenaya and his Ahwahneechee were eventually captured and their village burned; they were removed to a reservation near Fresno, California. The chief and some others were later allowed to return to Yosemite Valley. In the spring of 1852 they attacked a group of eight gold miners, and then moved east to flee law enforcement. Near Mono Lake, they took refuge with the nearby Mono tribe of Paiute. They stole horses from their hosts and moved away, but the Mono Paiutes tracked down and killed many of the Ahwahneechee, including Chief Tenaya. The Mono Paiute took the survivors as captives back to Mono Lake and absorbed them into the Mono Lake Paiute tribe.

After these wars, a number of Native Americans continued to live within the boundaries of Yosemite. A number of Indians supported the growing tourism industry by working as laborers or maids. Later, Indians became part of the tourism industry itself by selling baskets or performing for tourists. A reconstructed "Indian Village of Ahwahnee" has been erected behind the Yosemite Museum, located next to the Yosemite Valley Visitor Center.

The Dead Giant.

Vernal Fall.

In 1855, entrepreneur James Mason Hutchings, artist Thomas Ayres and two others were the first to tour the area. Hutchings and Ayres were responsible for much of the earliest publicity about Yosemite, writing articles and special magazine issues about the Valley. Ayres' style in art was highly detailed with exaggerated angularity. His works and written accounts were distributed nationally, and an art exhibition of his drawings was held in New York City. Hutchings' publicity efforts between 1855 and 1860 led to an increase in tourism to Yosemite.

Wawona was an Indian encampment in what is now the southwestern part of the park. Settler Galen Clark discovered the Mariposa Grove of giant sequoia in Wawona in 1857. He had simple lodgings built, and roads to the area. In 1879, the Wawona Hotel was built to serve tourists visiting Mariposa Grove. As tourism increased, so did the number of trails and hotels developed by people intending to build on the trade.

The Wawona Tree, also known as the Tunnel Tree, was a famous giant sequoia that stood in the Mariposa Grove. It was 227 feet (69 m) tall, and was 90 ft (27 m) in circumference. When a carriage-wide tunnel was cut through the tree in 1881, it became even more popular as a tourist photo attraction. Everything from horse-drawn carriages in the late 19th century, to automobiles in the first part of the 20th century, traveled the road which passed through that tree. The Wawona Tree fell in 1969 under a heavy load of snow. It was estimated to have been 2,300 years old.

Yosemite's first concession was established in 1884 when John Degnan and his wife established a bakery and store. In 1916, the National Park Service granted a 20-year concession to the Desmond Park Service Company. It bought out or built hotels, stores, camps, a dairy, a garage, and other park services. Desmond changed its name to the Yosemite National Park Company in December 1917 and was reorganized in 1920.

The Curry Company had been started in 1899 by David and Jennie Curry to provide concessions in the park. They also founded Camp Curry, formerly known as Half Dome Village, now restored back to Curry Village. The Currys lobbied reluctant park supervisors to allow expansion of concession operations and development in the area.

Administrators in the National Park Service felt that limiting the number of concessionaires in each national park would be more financially sound. The Curry Company and its rival, the Yosemite National Park Company, were forced to merge in 1925 to form the Yosemite Park and Curry Company (YP&CC). The company built the Ahwahnee Hotel in 1927.

Yosemite Grant

Concerned by the effects of commercial interests, prominent citizens including Galen Clark and Senator John Conness advocated for protection of the area. A park bill was prepared with the assistance of the General Land Office in the Interior Department. The bill passed both houses of the 38th United States Congress, and was signed by President Abraham Lincoln on June 30, 1864, creating the Yosemite Grant. This is the first instance of park land being set aside specifically for preservation and public use by action of the U.S. federal government, and set a precedent for the 1872 creation of Yellowstone as the first national park. Yosemite Valley and the Mariposa Grove were ceded to California as a state park, and a board of commissioners was proclaimed two years later.

Galen Clark was appointed by the commission as the Grant's first guardian, but neither Clark nor the commissioners had the authority to evict homesteaders (which included Hutchings). The issue was not settled until 1872 when the homesteader land holdings were invalidated by the U.S. Supreme Court. Clark and the reigning commissioners were ousted in 1880, this dispute also reaching the Supreme Court in 1880. The two Supreme Court decisions affecting management of the Yosemite Grant are considered important precedents in land management law. Hutchings became the new park guardian.

Access to the park by tourists improved in the early years of the park, and conditions in the Valley were made more hospitable. Tourism significantly increased after the First Transcontinental Railroad was completed in 1869, but the long horseback ride to reach the area was a deterrent. Three stagecoach roads were built in the mid-1870s to provide better access for the growing number of visitors to Yosemite Valley.

John Muir was a Scottish-born American naturalist and explorer. It was because of Muir that many National Parks were left untouched, such as Yosemite Valley National Park. One of the most significant camping trips Muir took was in 1903 with then president Theodore Roosevelt. This trip persuaded Roosevelt to return "Yosemite Valley and Mariposa Grove to federal protection as part of Yosemite National Park".

Muir wrote articles popularizing the area and increasing scientific interest in it. Muir was one of the first to theorize that the major landforms in Yosemite Valley were created by large alpine glaciers, bucking established scientists such as Josiah Whitney, who regarded Muir as an amateur. Muir wrote scientific papers on the area's biology. Landscape architect Frederick Law Olmsted emphasized the importance of conservation of Yosemite Valley.

Increased Protection Efforts

Overgrazing of meadows (especially by sheep), logging of giant sequoia, and other damage caused Muir to become an advocate for further protection. Muir convinced prominent guests of the importance of putting the area under federal protection; one such guest was Robert Underwood Johnson, editor of *Century Magazine.* Muir and Johnson lobbied Congress for the Act that created Yosemite National Park on October 1, 1890. The State of California, however, retained control of Yosemite Valley and Mariposa Grove. Muir also helped persuade local officials to virtually eliminate grazing from the Yosemite high country.

The newly created national park came under the jurisdiction of the United States Army's Troop I of the 4th Cavalry on May 19, 1891, which set up camp in Wawona with Captain Abram Epperson Wood as acting superintendent. By the late 1890s, sheep grazing was no longer a problem, and the Army made other improvements. The cavalry could not intervene to ease the worsening condition of Yosemite Valley and Mariposa Grove. The cavalry left another legacy in the park, the ranger hat. From 1899 to 1913, cavalry regiments of the Western Department, including the all Black 9th Cavalry (known as the "Buffalo Soldiers") and the 1st Cavalry, stationed two troops at Yosemite and brought with them the trooper's campaign hat with its distinctive Montana Peak we recognize today as the "ranger hat". This peak had been formed into the trooper's stetson by veterans of the 1898 Spanish–American War to better shed tropical rain.

Muir and his Sierra Club continued to lobby the government and influential people for the creation of a unified Yosemite National Park. In May 1903, President Theodore Roosevelt camped with Muir near Glacier Point for three days. On that trip, Muir convinced Roosevelt to take control of Yosemite Valley and Mariposa Grove away from California and return it to the federal government. In 1906, Roosevelt signed a bill that did precisely that.

National Park Service

The National Park Service was formed in 1916, and Yosemite was transferred to that agency's jurisdiction. Tuolumne Meadows Lodge, Tioga Pass Road, and campgrounds at Tenaya and Merced lakes were also completed in 1916. Automobiles started to enter the park in ever-increasing numbers following the construction of all-weather highways to the park. The Yosemite Museum was founded in 1926 through the efforts of Ansel Franklin Hall.In the 1920s, the museum featured Native Americans practicing traditional crafts, and many of the Sierra Miwok continued to live in Yosemite Valley until they were evicted from Yosemite in the 1960s.

In 1903, a dam in the northern portion of the park was proposed. Located in the Hetch Hetchy Valley, its purpose was to provide water and hydroelectric power to San Francisco. Muir and the Sierra Club opposed the project, while others, including Gifford Pinchot, supported it. In 1913, the U.S. Congress authorized the O'Shaughnessy Dam through passage of the Raker Act.

O'Shaughnessy Dam in Hetch Hetchy Valley.

In the late 1920s a bid for Yosemite for the 1932 Winter Olympics was put forward. Ultimately, the 1932 Winter Olympics were awarded to Lake Placid, New York. In 1937, conservationist Rosalie Edge, head of the Emergency Conservation Committee (ECC), successfully lobbied Congress to purchase about 8,000 acres of old-growth sugar pines on the perimeter of Yosemite National Park that were to be logged.

More recently, preservationists persuaded Congress to designate 677,600 acres (274,200 ha), or about 89% of the park, as the Yosemite Wilderness—a highly protected wilderness area. The Park Service has reduced artificial inducements to visit the park, such as the Firefall, in which red-hot embers were pushed off a cliff near Glacier Point at night. Traffic congestion in Yosemite Valley during the summer months has become a concern. Two electric buses commenced service in September 1995. The buses are quiet and do not emit pollutants. Eventually, all the buses in Yosemite will be electric.

In 2016, The Trust for Public Land purchased Ackerson Meadow, a 400-acre tract on the western edge of Yosemite National Park, for $2.3 million in order to preserve habitat and protect the area from development. Ackerson Meadow was originally included in the proposed 1890 park boundary but never acquired by the federal government. On September 7, 2016, the National Park Service accepted the donation of the land, making the meadow the largest addition to Yosemite since 1949.

Geography

Yosemite National Park is located in the central Sierra Nevada of California. Three wilderness areas are adjacent to Yosemite: the Ansel Adams Wilderness to the southeast, the Hoover Wilderness to the northeast, and the Emigrant Wilderness to the north.

The 1,189 sq mi (3,080 km²) park is roughly the size of the U.S. state of Rhode Island and contains thousands of lakes and ponds, 1,600 miles (2,600 km) of streams, 800 miles (1,300 km) of hiking trails, and 350 miles (560 km) of roads. Two federally designated Wild and Scenic Rivers, the Merced and the Tuolumne, begin within Yosemite's borders and flow westward through the Sierra foothills, into the Central Valley of California. On average, about 4 million people visit the park each year, with most visitor use concentrated in the seven-square-mile (18 km²) area of Yosemite Valley.

Rocks and Erosion

El Capitan, a granite monolith on Yosemite Valley's northern escarpment.

Almost all of the landforms in the Yosemite area are cut from the granitic rock of the Sierra Nevada Batholith (a batholith is a large mass of intrusive igneous rock that formed deep below the surface). About 5% of the park's landforms (mostly in its eastern margin near Mount Dana) are metamorphosed volcanic and sedimentary rocks. These rocks are called *roof pendants* because they were once the roof of the underlying granitic rock.

Erosion acting upon different types of uplift-created joint and fracture systems is responsible for creating the valleys, canyons, domes, and other features we see today. These joints and fracture systems do not move, and are therefore not faults. Spacing between joints is controlled by the amount of silica in the granite and granodiorite rocks; more silica tends to create a more resistant rock, resulting in larger spaces between joints and fractures.

Pillars and columns, such as Washington Column and Lost Arrow, are created by cross joints. Erosion acting on master joints is responsible for creating valleys and later canyons. The single most erosive force over the last few million years has been large alpine glaciers, which have turned the previously V-shaped river-cut valleys into U-shaped glacial-cut canyons (such as Yosemite Valley and Hetch Hetchy Valley). Exfoliation (caused by the tendency of crystals in plutonic rocks to expand at the surface) acting on granitic rock with widely spaced joints is responsible for creating domes such as Half Dome and North Dome and inset arches like Royal Arches.

Popular Features

Yosemite Valley represents only one percent of the park area, but this is where most visitors arrive and stay. The Tunnel View is the first view of the Valley for many visitors and is extensively photographed. El Capitan, a prominent granite cliff that looms over Yosemite Valley, is one of the most popular rock climbing destinations in the world because of its diverse range of climbing routes in addition to its year-round accessibility. Granite domes such as Sentinel Dome and Half Dome rise 3,000 and 4,800 feet (910 and 1,460 m), respectively, above the valley floor. In the park are many domes.

Cathedral Peak.

The high country of Yosemite contains beautiful areas such as Tuolumne Meadows, Dana Meadows, the Clark Range, the Cathedral Range, and the Kuna Crest. The Sierra crest and the Pacific Crest Trail run through Yosemite, with peaks of red metamorphic rock, such as Mount Dana and Mount Gibbs, and granite peaks, such as Mount Conness. Mount Lyell is the highest point in the park, standing at 13,120 feet (4,000 m). The Lyell Glacier is the largest glacier in Yosemite National Park and is one of the few remaining in the Sierra Nevada today.

The park has three groves of ancient giant sequoia (*Sequoiadendron giganteum*) trees; the Mariposa Grove (200 trees), the Tuolumne Grove (25 trees), and the Merced Grove (20 trees). This species grows larger in volume than any other and is one of the tallest and longest-lived.

Water and Ice

The Tuolumne and Merced River systems originate along the crest of the Sierra Nevada in the park and have carved river canyons 3,000 to 4,000 feet (910 to 1,220 m) deep. The Tuolumne River drains the entire northern portion of the park, an area of approximately 680 square miles (1,800 km^2). The Merced River begins in the park's southern peaks, primarily the Cathedral and Clark Ranges, and drains an area of approximately 511 square miles (1,320 km²).

The Merced River flowing through Yosemite Valley, a U-shaped valley.

Hydrologic processes, including glaciation, flooding, and fluvial geomorphic response, have been fundamental in creating landforms in the park. The park also contains approximately 3,200 lakes (greater than 100 m²), two reservoirs, and 1,700 miles (2,700 km) of streams, all of which help form these two large watersheds. Wetlands in Yosemite occur in valley bottoms throughout the park, and are often hydrologically linked to nearby lakes and rivers through seasonal flooding and groundwater movement. Meadow habitats, distributed at elevations from 3,000 to 11,000 feet (910 to 3,350 m) in the park, are generally wetlands, as are the riparian habitats found on the banks of Yosemite's numerous streams and rivers.

Bridalveil Fall flows from a U-shaped hanging valley that was created by a tributary glacier.

Yosemite is famous for its high concentration of waterfalls in a small area. Numerous sheer drops, glacial steps and hanging valleys in the park provide many places for waterfalls to exist, especially during April, May, and June (the snowmelt season). Located in Yosemite Valley, the Yosemite Falls is the highest in North America at 2,425 feet (739 m). Also in Yosemite Valley is the much lower volume Ribbon Falls, which has the highest single vertical drop, 1,612 feet (491 m). Perhaps the most prominent of the Yosemite Valley waterfalls is Bridalveil Fall, which is the waterfall seen from the Tunnel View viewpoint at the east end of the Wawona Tunnel. Wapama Falls in Hetch Hetchy Valley is another notable waterfall. Hundreds of ephemeral waterfalls also exist in the park.

All glaciers in the park are relatively small glaciers that occupy areas that are in almost permanent shade, such as north- and northeast-facing cirques. Lyell Glacier is the largest glacier in Yosemite (the Palisades Glaciers are the largest in the Sierra Nevada) and covers 160 acres (65 ha). None of the Yosemite glaciers are a remnant of the much, much larger Ice Age alpine glaciers responsible for sculpting the Yosemite landscape. Instead, they were formed during one of the neoglacial episodes that have occurred since the thawing of the Ice Age (such as the Little Ice Age). Climate change has reduced the number and size of glaciers around the world. Many Yosemite glaciers, including Merced Glacier, which was discovered by John Muir in 1871 and bolstered his glacial origins theory of the Yosemite area, have disappeared and most of the others have lost up to 75% of their surface area.

Climate

Yosemite in autumn.

Yosemite has a Mediterranean climate (Köppen climate classification *Csa*), meaning most precipitation falls during the mild winter, and the other seasons are nearly dry (less than 3% of precipitation falls during the long, hot summers). Because of orographic lift, precipitation increases with elevation up to 8,000 feet (2,400 m) where it slowly decreases to the crest. Precipitation amounts vary from 36 inches (910 mm) at 4,000 feet (1,200 m) elevation to 50 inches (1,300 mm) at 8,600 feet (2,600 m). Snow does not typically persist on the ground until November in the high country. It accumulates all winter and into March or early April.

Mean daily temperatures range from 25 °F (−4 °C) to 53 °F (12 °C) at Tuolumne Meadows at 8,600 feet (2,600 m). At the Wawona Entrance (elevation 5,130 feet or 1,560 metres), mean daily temperature ranges from 36 to 67 °F (2 to 19 °C). At the lower elevations below 5,000 feet (1,500 m), temperatures are hotter; the mean daily high temperature at Yosemite Valley (elevation 3,966 feet or 1,209 metres) varies from 46 to 90 °F (8 to 32 °C). At elevations above 8,000 feet (2,400 m), the hot, dry summer temperatures are moderated by frequent summer thunderstorms, along with snow that can persist into July. The combination of dry vegetation, low relative humidity, and thunderstorms results in frequent lightning-caused fires as well.

At the park headquarters, with an elevation of 3,966 feet (1,209 m), January averages 38.2 °F (3.4 °C), while July averages 73.0 °F (22.8 °C), though in summer the nights are much cooler than the hot days. There are an average of 39.5 days with highs of 90 °F (32 °C) or higher and an average of 97.9 nights with freezing temperatures. Freezing temperatures have been recorded in every month of the year. The record high temperature was 115 °F (46 °C) on July 20, 1915, while the record low temperature was −6 °F (−21 °C) on January 2, 1924 and on January 21, 1937. Average annual precipitation is nearly 37 inches (940 mm), falling on 65 days. The wettest year was 1983 with 68.94 inches (1,751 mm) and the driest year was 1976 with 14.84 inches (377 mm). The most precipitation in one month was 29.61 inches (752 mm) in December 1955 and the most in one day was 6.92 inches (176 mm) on December 23, 1955. Average annual snowfall is 65.6 inches (1.67 m). The snowiest year was 1967 with 154.9 inches (3.93 m). The most snow in one month was 140.8 inches (3.58 m) in January 1993.

Geology

Tectonic and Volcanic Activity

The area of the park was astride a passive continental margin during the Precambrian and early Paleozoic. Sediment was derived from continental sources and was deposited in shallow water. These rocks have since been metamorphosed.

Heat generated from the Farallon Plate subducting below the North American Plate led to the creation of an island arc of volcanoes on the west coast of proto-North America between the late Devonian and Permian periods. Later volcanism in the Jurassic intruded and covered these rocks in what may have been magmatic activity associated with the early stages of the creation of the Sierra Nevada Batholith. 95% of these rocks were eventually removed by uplifted-accelerated erosion.

The first phase of regional plutonism started 210 million years ago in the late Triassic and continued throughout the Jurassic to about 150 million years before present (BP). Around the same time, the Nevadan orogeny built the Nevadan mountain range (also called the Ancestral Sierra Nevada) to a height of 15,000 feet (4,600 m). This was directly part of the creation of the Sierra Nevada Batholith, and the resulting rocks were mostly granitic in composition and emplaced about 6 miles (9.7 km) below the surface. The second major pluton emplacement phase lasted from about 120 million to 80 million years ago during the Cretaceous. This was part of the Sevier orogeny.

Starting 20 million years ago (in the Cenozoic) and lasting until 5 million years ago, a now-extinct extension of Cascade Range volcanoes erupted, bringing large amounts of igneous material in the area. These igneous deposits blanketed the region north of the Yosemite region. Volcanic activity persisted past 5 million years BP east of the current park borders in the Mono Lake and Long Valley areas.

Uplift and Erosion

Exfoliation joints cause erosion in granitic rocks, creating many domes including Half Dome.

Starting 10 million years ago, vertical movement along the Sierra fault started to uplift the Sierra Nevada. Subsequent tilting of the Sierra block and the resulting accelerated uplift of the Sierra Nevada increased the gradient of western-flowing streams. The streams consequently ran faster and thus cut their valleys more quickly. Additional uplift occurred when major faults developed to the east, especially the creation of Owens Valley from Basin and Range-associated extensional forces. Uplift of the Sierra accelerated again about two million years ago during the Pleistocene.

The uplifting and increased erosion exposed granitic rocks in the area to surface pressures, resulting in exfoliation (responsible for the rounded shape of the many domes in the park) and mass wasting following the numerous fracture joint planes (cracks; especially vertical ones) in the now solidified plutons. Pleistocene glaciers further accelerated this process and the larger ones transported the resulting talus and till from valley floors.

Numerous vertical joint planes controlled where and how fast erosion took place. Most of these long, linear and very deep cracks trend northeast or northwest and form parallel, often regularly spaced sets. They were created by uplift-associated pressure release and by the unloading of overlying rock via erosion.

Sculpting by Glaciers

A series of glaciations further modified the region starting about 2 to 3 million years ago and ending sometime around 10,000 BP. At least four major glaciations have occurred in the Sierra Nevada, locally called the Sherwin (also called the pre-Tahoe), Tahoe,

Tenaya, and Tioga. The Sherwin glaciers were the largest, filling Yosemite and other valleys, while later stages produced much smaller glaciers. A Sherwin-age glacier was almost surely responsible for the major excavation and shaping of Yosemite Valley and other canyons in the area.

Glacially polished granite cirque in upper Tenaya Canyon.

Glacial systems reached depths of up to 4,000 feet (1,200 m) and left their marks in the Yosemite area. The longest glacier in the Yosemite area ran down the Grand Canyon of the Tuolumne River for 60 miles (97 km), passing well beyond Hetch Hetchy Valley. Merced Glacier flowed out of Yosemite Valley and into the Merced River Gorge. Lee Vining Glacier carved Lee Vining Canyon and emptied into Lake Russel (the much-enlarged ice age version of Mono Lake). Only the highest peaks, such as Mount Dana and Mount Conness, were not covered by glaciers. Retreating glaciers often left recessional moraines that impounded lakes such as the 5.5 miles (9 km) long Lake Yosemite (a shallow lake that periodically covered much of the floor of Yosemite Valley).

Ecology

Habitats

Mule deer in Yosemite Valley.

With its scrubby sun-baked chaparral, stately groves of pine, fir, and sequoia, and expanses of alpine woodlands and meadows, Yosemite National Park preserves a Sierra

Nevada landscape as it prevailed before Euro-American settlement. In contrast to surrounding lands, which have been significantly altered by logging, the park still contains some 225,510 acres (91,260 ha) of old-growth forest. Taken together, the park's varied habitats support over 250 species of vertebrates, which include fish, amphibians, reptiles, birds, and mammals.

Much of Yosemite's western boundary has habitats dominated by mixed coniferous forests of ponderosa pine, sugar pine, incense cedar, white fir, Douglas fir, and a few stands of giant sequoia, interspersed by areas of black oak and canyon live oak. A relatively high diversity of wildlife species is supported by these habitats, because of relatively mild, lower-elevation climate and the mixture of habitat types and plant species. Wildlife species typically found in these habitats include black bear, coyote, raccoon, mountain kingsnake, Gilbert's skink, white-headed woodpecker, bobcat, river otter, gray fox, red fox, brown creeper, two species of skunk, cougar, spotted owl, and a wide variety of bat species.

Going higher in elevation, the coniferous forests become purer stands of red fir, western white pine, Jeffrey pine, lodgepole pine, and the occasional foxtail pine. Fewer wildlife species tend to be found in these habitats, because of their higher elevation and lower complexity. Species likely to be found include golden-mantled ground squirrel, chickaree, fisher, Steller's jay, hermit thrush, and northern goshawk. Reptiles are not common, but include rubber boa, western fence lizard, and northern alligator lizard.

Marmot in Tuolumne Meadows.

As the landscape rises, trees become smaller and more sparse, with stands broken by areas of exposed granite. These include lodgepole pine, whitebark pine, and mountain hemlock that, at highest elevations, give way to vast expanses of granite as treeline is reached. The climate in these habitats is harsh and the growing season is short, but species such as pika, yellow-bellied marmot, white-tailed jackrabbit, Clark's nutcracker, and black rosy finch are adapted to these conditions. Also, the treeless alpine habitats are the areas favored by Sierra Nevada bighorn sheep. This species, however, is now found in the Yosemite area only around Tioga Pass, where a small, reintroduced population exists.

At a variety of elevations, meadows provide important, productive habitat for wildlife. Animals come to feed on the green grasses and use the flowing and standing water found in many meadows. Predators, in turn, are attracted to these areas. The interface between meadow and forest is also favored by many animal species because of the proximity of open areas for foraging and cover for protection. Species that are highly dependent upon meadow habitat include great grey owl, willow flycatcher, Yosemite toad, and mountain beaver.

Management Issues

A black bear with an ear tag in Yosemite Valley.

The black bears of Yosemite were once famous for breaking into parked cars to steal food. They were also an encouraged tourist sight for many years at the park's garbage dumps, where bears congregated to eat park visitors' garbage and tourists gathered to photograph the bears. Increasing encounters between bears and humans and increasing damage to property led to an aggressive campaign to discourage bears from relying on human food or interacting with people and their property. The open-air dumps were closed; all trash receptacles were replaced with bear-proof receptacles; all campgrounds were equipped with bear-proof food lockers so that people would not leave food in their vehicles, which were easy targets for the powerful and resourceful bears. Because bears who show aggression towards people usually are eventually destroyed, park personnel have continued to come up with innovative ways to have bears associate humans and their property with unpleasant experiences, such as being hit with rubber bullets. Today, about 30 bears a year are captured and ear-tagged and their DNA is sampled so that, when bear damage occurs, rangers can ascertain which bear is causing the problem.

Despite the richness of high-quality habitats in Yosemite, the brown bear, California condor, and least Bell's vireo have become extinct in the park within historical time, and another 37 species currently have special status under either California or federal endangered species legislation. The most serious current threats to Yosemite's wildlife and the ecosystems they occupy include loss of a natural fire regime, exotic species, air pollution, habitat fragmentation, and climate change. On a more local basis, factors such as road kills and the availability of human food have affected some wildlife species.

Yosemite National Park has documented more than 130 non-native plant species within park boundaries. These non-native plants were introduced into Yosemite following the migration of early Euro-American settlers in the late 1850s. Natural and human-caused disturbances, such as wildland fires and construction activities, have contributed to a rapid increase in the spread of non-native plants. A number of these species aggressively invade and displace the native plant communities, resulting in impacts on the park's resources. Non-native plants can bring about significant changes in park ecosystems by altering the native plant communities and the processes that support them. Some non-native species may cause an increase in the fire frequency of an area or increase the available nitrogen in the soil that may allow more non-native plants to become established. Many non-native species, such as yellow star thistle (Centaurea solstitialis), are able to produce a long tap root that allows them to out-compete the native plants for available water.

The yellow star thistle competes with Yosemite's native plants.

Bull thistle (Cirsium vulgare), common mullein (Verbascum thapsus), and Klamath weed (Hypericum perforatum) have been identified as noxious pests in Yosemite since the 1940s. Additional species that have been recognized more recently as aggressive and requiring control are yellow star thistle (Centaurea solstitialis), sweet clover (Melilot spp.), Himalayan blackberry (Rubus armeniacus), cut-leaved blackberry (Rubus laciniatus) and large periwinkle (Vinca major).

Increasing ozone pollution is causing tissue damage to the massive giant sequoia trees in the park, making them more vulnerable to insect infestation and disease. Since the cones of these trees require fire-touched soil to germinate, historic fire suppression has reduced these trees' ability to reproduce. The current policy of setting prescribed fires is expected to help the germination issue.

Wildfires

Forest fires seasonally clear the park of dead vegetation, making way for new growth. These fires damage the income generated by tourism. The Rim Fire in 2013 destroyed nearly $2 billion in assets and revenue, though natural woodland assets are renewable, and closed off much of the park to tourists. This fire was the third largest on record, and burned nearly 500 acres of wild habitat.

During late July and early August, 2018, sections of the park, including the Valley, were temporarily closed due to the Ferguson Fire at its western boundary. The closing was the largest in almost thirty years at the park.

The Meadow Fire burns in Little Yosemite Valley.

NATURAL MONUMENT OR FEATURE

Natural monument or feature is a protected area set aside to protect a specific natural monument, which can be a landform, sea mount, and submarine cavern, geological feature such as a cave or even a living feature such as an ancient grove. They are generally quite small protected areas and often have high visitor value.

Primary Objective

To protect specific outstanding natural features and their associated biodiversity and habitats.

Other Objectives

- To provide biodiversity protection in landscapes or seascapes that have otherwise undergone major changes.
- To protect specific natural sites with spiritual and cultural values where these also have biodiversity values.
- To conserve traditional spiritual and cultural values of the site.

Distinguishing Features

Category III protected areas are usually relatively small sites that focus on one or more prominent natural features and the associated ecology, rather than on a broader ecosystem. They are managed in much the same way as category II. The term "natural" as used here can refer to both wholly natural features (the commonest use) but also sometimes features that have been influenced by humans. In the latter case these sites should also always have important associated biodiversity attributes, which should be reflected as a priority in their management objectives if they are to be classified as a protected area rather than an historical or spiritual site. Category III protected areas could include:

- Natural geological and geomorphological features such as waterfalls, cliffs, craters, caves, fossil beds, sand dunes, rock forms, valleys and marine features such as sea mounts or coral formations.
- Culturally-influenced natural features such as cave dwellings and ancient tracks.
- Natural-cultural sites: such as the many forms of sacred natural sites (sacred groves, springs, waterfalls, mountains, sea coves etc.) of importance to one or more faith groups.
- Cultural sites with associated ecology where protection of a cultural site also protects significant and important biodiversity, such as archaeological/historical sites that are inextricably linked to a natural area.

Nature conservation attributes of category III protected areas fall into two main types:

- Biodiversity that is uniquely related to the ecological conditions associated with the natural feature – such as the spray zones of a waterfall, the ecological conditions in caves or plant species confined to cliffs.
- Biodiversity that is surviving because the presence of cultural or spiritual values at the site have maintained a natural or semi-natural habitat in what is otherwise a modified ecosystem – such as some sacred natural sites or historical sites that have associated natural areas. In these cases the key criteria for inclusion as a protected area will be (i) value of the site as a contribution to broad-scale conservation and (ii) prioritization of biodiversity conservation within management plans.

Category III has been suggested as providing a natural management approach for many sacred natural sites, such as sacred groves. Although sacred natural sites are found in all categories and can benefit from a wide range of management approaches, they may be particularly suited to management as natural monuments.

Role in the Landscape/Seascape

Category III is really intended to protect the unusual rather than to provide logical components in a broad-scale approach to conservation, so that their role in landscape or ecoregional strategies may sometimes be opportunistic rather than planned. In other cases (e.g., cave systems) such sites may play a key ecological role identified within wider conservation plans.

Important natural monuments can sometimes provide an incentive for protection and anopportunity for environmental/cultural education even in areas where other forms of protection are resisted due to population or development pressure, such as important sacred or cultural sites and in these cases category III can preserve samples of natural habitat in otherwise cultural or fragmented landscapes.

What makes Category III Unique?

Because it is aimed at protecting a particular feature, category III is perhaps the most heavily influenced of all the categories by human perceptions of what is of value in a landscape or seascape rather than by any more quantitative assessments of value. This is less applicable in category III protected areas designated for geological features, where systematic identification is possible. Management is usually focused on protecting and maintaining particular natural features.

The fact that an area contains an important natural monument does not mean that it will inevitably be managed as a category III; for instance the Grand Canyon in Arizona is managed as category II, despite being one of the most famous natural monuments in the world, because it is also a large and diverse area with associated recreation activities making it better suited to a category II model. Category III is most suitable where the protection of the feature is the sole or dominant objective.

Category III differs from the other categories in the following ways:

Category Ia Category Ib	Category III is not confined to natural and pristine landscapes but could be established in areas that are otherwise cultural or fragmented landscapes. Visitation and recreation is often encouraged and research and monitoring limited to the understanding and maintenance of a particular natural feature.
Category II	The emphasis of category III management is not on protection of the whole ecosystem, but of particular natural features; otherwise category III is similar to category II and managed in much the same way but at a rather smaller scale in both size and complexity of management.
Category IV	The emphasis of category III management is not on protection of the key species or habitats, but of particular natural features.
Category V	Category III is not confined to cultural landscapes and management practices will probably focus more on stricter protection of the particular feature than in the case of category V.
Category VI	Category III is not aimed at sustainable resource use.

Issues for Consideration

- It will sometimes be difficult to ascertain the conservation attributes of category III sites, particularly in cases where there may be pressure to accept sites within a protected area system to help protect cultural or spiritual values.

- Not all natural monuments are permanent – while some sacred trees have survived for a thousand years or more they will eventually die – indeed many trees are considered to be sacred in part because they are already very old. It is not clear what happens to a category III protected area if its key natural monument dies or degrades.
- It is sometimes difficult to draw the boundaries between a natural monument and cultural site, particularly where archaeological remains are included within category III.
- Some apparent "monuments" may require protection of a larger ecosystem to survive – for example a waterfall may require protection of a whole watershed to maintain its flow.

HABITAT/SPECIES MANAGEMENT AREA

Habitat/Species Management Area is a protected area aiming to protect particular species or habitats and management reflects this priority. Many category IV protected areas will need regular, active interventions to address the requirements of particular species or to maintain habitats, but this is not a requirement of the category.

Primary Objective

To maintain, conserve and restore species and habitats.

Other Objectives

- To protect vegetation patterns or other biological features through traditional management approaches.
- To protect fragments of habitats as components of landscape or seascape-scale conservation strategies.

- To develop public education and appreciation of the species and habitats concerned.
- To provide a means by which the urban residents may obtain regular contact with nature.

Distinguishing Features

Category IV protected areas usually help to protect, or restore: 1) flora species of international, national or local importance; 2) fauna species of international, national or local importance including resident or migratory fauna; and 3) habitats. The size of the area varies but can often be relatively small; this is however not a distinguishing feature. Management will differ depending on need. Protection may be sufficient to maintain particular habitats and species. However, as category IV protected areas often include fragments of an ecosystem, these areas may not be self-sustaining and will require regular and active management interventions to ensure the survival of specific habitats and to meet the requirements of particular species. A number of approaches are suitable:

- Protection of particular species: To protect particular target species, which will usually be under threat (e.g., one of the last remaining populations).
- Protection of habitats: To maintain or restore habitats, which will often be fragments of ecosystems.
- Active management to maintain target species: To maintain viable populations of particular species, which might include for example artificial habitat creation or maintenance (such as artificial reef creation), supplementary feeding or other active management systems.
- Active management of natural or semi-natural ecosystems: To maintain natural or semi-natural habitats that are either too small or too profoundly altered to be self-sustaining, e.g., if natural herbivores are absent they may need to be replaced by livestock or manual cutting; or if hydrology has been altered this may necessitate artificial drainage or irrigation.
- Active management of culturally-defined ecosystems: To maintain cultural management systems where these have a unique associated biodiversity. Continual intervention is needed because the ecosystem has been created or at least substantially modified by management. The primary aim of management is maintenance of associated biodiversity.

Active management means that the overall functioning of the ecosystem is being modified by e.g., halting natural succession, providing supplementary food or artificially creating habitats: i.e., management will often include much more than just addressing threats, such as poaching or invasive species, as these activities take place in virtually all protected areas in any category and are therefore not diagnostic. Category IV protected areas will generally be publicly accessible.

Role in the Landscape/Seascape

Category IV protected areas frequently play a role in "plugging the gaps" in conservation strategies by protecting key species or habitats in ecosystems. They could, for instance, be used to:

- Protect critically endangered populations of species that need particular management interventions to ensure their continued survival.
- Protect rare or threatened habitats including fragments of habitats.
- Secure stepping-stones (places for migratory species to feed and rest) or breeding sites.
- Provide flexible management strategies and options in buffer zones around, or connectivity conservation corridors between, more strictly protected areas that are more acceptable to local communities and other stakeholders.
- Maintain species that have become dependent on cultural landscapes where their original habitats have disappeared or been altered.

What makes Category IV Unique?

Category IV provides a management approach used in areas that have already undergone substantial modification, necessitating protection of remaining fragments, with or without intervention.

Category IV differs from the other categories in the following ways:

Category Ia	Category IV protected areas are not strictly protected from human use; scientific research may take place but generally as a secondary objective.
Category Ib	Category IV protected areas cannot be described as "wilderness", as defined by IUCN. Many will be subject to management intervention that is inimical to the concept of category Ib wilderness areas; those that remain un-managed are likely to be too small to fulfil the aims of category Ib.
Category II	Category IV protected areas aim their conservation at particular species or habitats and may in consequence have to pay less attention to other elements of the ecosystem in consequence, whereas category II protected areas aim to conserve fully functional ecosystems. Categories II and IV may in some circumstances closely resemble each other and the distinction is partly a matter of objective – i.e., whether the aim is to protect to the extent possible the entire ecosystem (category II) or is focused to protect a few key species or habitats (category IV).

Category III	The objective of category IV areas is of a more biological nature whereas category III is site-specific and more morphologically or culturally oriented.
Category V	Category IV protected areas aim to protect identified target species and habitats whereas category V aims to protect overall landscapes/seascapes with value for nature conservation. Category V protected areas will generally possess socio-cultural characteristics that may be absent in IV. Where category IV areas may use traditional management approaches this will explicitly be to maintain associated species as part of a management plan and not more broadly as part of a management approach that includes a wide range of for-profit activities.
Category VI	Management interventions in category IV protected areas are primarily aimed at maintaining species or habitats while in category VI protected areas they are aimed at linking nature conservation with the sustainable use of resources. As with category V, category VI protected areas are generally larger than category IV.

Issues for Consideration

- Many category IV protected areas exist in crowded landscapes and seascapes, where human pressure is comparatively greater, both in terms of potential illegal use and visitor pressure.
- The category IV protected areas that rely on regular management intervention need appropriate resources from the management authority and can be relatively expensive to maintain unless management is undertaken voluntarily by local communities or other actors.
- Because they usually protect part of an ecosystem, successful long-term management of category IV protected areas necessitates careful monitoring and an even greater-than-usual emphasis on overall ecosystem approaches and compatible management in other parts of the landscape or seascape.

PROTECTED LANDSCAPE/SEASCAPE AREA

Protected landscape/seascape is a protected area where the interaction of people and nature over time has produced an area of distinct character with significant ecological, biological, cultural and scenic value: and where safeguarding the integrity of this interaction is vital to protecting and sustaining the area and its associated nature conservation and other values.

Primary Objective

To protect and sustain important landscapes/seascapes and the associated nature conservation and other values created by interactions with humans through traditional management practices.

Other Objectives

- To maintain a balanced interaction of nature and culture through the protection of landscape and seascape and associated traditional management approaches, societies, cultures and spiritual values.
- To contribute to broad-scale conservation by maintaining species associated with cultural landscapes and by providing conservation opportunities in heavily used landscapes.
- To provide opportunities for enjoyment, well-being and socio-economic activity through recreation and tourism.
- To provide natural products and environmental services.
- To provide a framework to underpin active involvement by the community in the management of valued landscapes or seascapes and the natural and cultural heritage that they contain.
- To encourage the conservation of agrobiodiversity and aquatic biodiversity.
- To act as models of sustainability so that lessons can be learnt for wider application.

Distinguishing Features

Category V protected areas result from biotic, abiotic and human interaction and should have the following essential characteristics:

- Landscape and coastal and island seascape of high and distinct scenic quality and with significant associated habitats, flora and fauna and associated cultural features.
- A balanced interaction between people and nature that has endured over time and still has integrity, or where there is reasonable hope of restoring that integrity.
- Unique or traditional land-use patterns, e.g., as evidenced in sustainable agricultural and forestry systems and human settlements that have evolved in balance with their landscape.

The following are desirable characteristics:

- Opportunities for recreation and tourism consistent with life style and economic activities.
- Unique or traditional social organizations, as evidenced in local customs, livelihoods and beliefs.
- Recognition by artists of all kinds and in cultural traditions (now and in the past).
- Potential for ecological and landscape restoration.

Role in the Landscape/Seascape

Generally, category V protected areas play an important role in conservation at the landscape/seascape scale, particularly as part of a mosaic of management patterns, protected area designations and other conservation mechanisms:

- Some category V protected areas act as a buffer around a core of one or more strictly protected areas to help to ensure that land and water-use activities do not threaten their integrity.
- Category V protected areas may also act as linking habitat between several other protected areas.

Category V offers unique contributions to conservation of biological diversity. In particular:

- Species or habitats that have evolved in association with cultural management systems and can only survive if those management systems are maintained.
- To provide a framework when conservation objectives need to be met over a large area (e.g., for top predators) in crowded landscapes with a range of ownership patterns, governance models and land use.
- In addition, traditional systems of management are often associated with important components of agrobiodiversity or aquatic biodiversity, which can be conserved only by maintaining those systems.

What makes Category V Unique?

Category V differs from the other categories in the following ways:

Category Ia	Human intervention is expected. Category V does not prioritize research, though it can offer opportunities to study interactions between people and nature.

Category Ib	Category V protected areas are not "wilderness" as defined by IUCN. Many will be subject to management intervention inimical to the concept of category Ib.
Category II	Category II seeks to minimize human activity in order to allow for "as natural a state as possible". Category V includes an option of continuous human interaction.
Category III	Category III focuses on specific features and single values and emphasises the monumentality, uniqueness and rarity of individual features, whereas these are not required for category V protected areas, which encompasses broader landscapes and multiple values.
Category IV	Category V aims to protect overall landscapes and seascapes that have value to biodiversity, whereas category IV aims often quite specifically to protect identified target species and habitats. Category V protected areas will often be larger than category IV.
Category VI	Category VI emphasises the need to link nature conservation in natural areas whilst supporting sustainable livelihoods: conversely category V emphasises values from long-term interactions of people and nature in modified conditions. In category VI the emphasis is on sustainable use of environmental products and services (typically hunting, grazing, management of natural resources), whereas in category V the emphasis is on more intensive uses (typically agriculture, forestry, tourism). Category VI will usually be more "natural" than category V.

Issues for Consideration

- Being a relatively flexible model, category V may sometimes offer conservation options where more strictly protected areas are not feasible.
- Category V protected areas can seek to maintain current practices, restore historical management systems or, perhaps most commonly, maintain key landscape values whilst accommodating contemporary development and change: decisions about this need to be made in management plans.
- The emphasis on interactions of people and nature over time raises the conceptual question for any individual category V protected area: at what point on the temporal continuum should management focus? And, in an area established to protect values based on traditional management systems, what happens when traditions change or are lost?
- Since social, economic and conservation considerations are all integral to the category V concept, defining measures of performance for all of these values is important in measuring success.
- As people are the stewards of the landscape or seascape in category V protected areas, clear guidelines are needed about the extent to which decision making can be left to local inhabitants and how far a wider public interest should prevail when there is conflict between local and national needs.

- How is category V distinguished from sustainable management in the wider landscape? As an area with exceptional values? As an example of best practice in management? Category V is perhaps the most quickly developing of any protected area management approaches.
- There are still only a few examples of the application of category V in coastal and marine settings where a "protected seascape" approach could be the most appropriate management option and more examples are needed (e.g., Holdaway undated).

PROTECTED AREA WITH SUSTAINABLE USE OF NATURAL RESOURCES

Protected area with sustainable use of natural resources is a protected area that conserve ecosystems and habitats, together with associated cultural values and traditional natural resource management systems. They are generally large, with most of the area in a natural condition, where a proportion is under sustainable natural resource management and where low-level non-industrial use of natural resources compatible with nature conservation is seen as one of the main aims of the area.

Primary Objective

To protect natural ecosystems and use natural resources sustainably, when conservation and sustainable use can be mutually beneficial.

Other Objectives

- To promote sustainable use of natural resources, considering ecological, economic and social dimensions.
- To promote social and economic benefits to local communities where relevant.
- To facilitate inter-generational security for local communities' livelihoods – therefore ensuring that such livelihoods are sustainable.

- To integrate other cultural approaches, belief systems and world-views within a range of social and economic approaches to nature conservation.
- To contribute to developing and maintaining a more balanced relationship between humans and the rest of nature.
- To contribute to sustainable development at national, regional and local level (in the last case mainly to local communities and indigenous peoples depending on the protected natural resources).
- To facilitate scientific research and environmental monitoring, mainly related to the conservation and sustainable use of natural resources.
- To collaborate in the delivery of benefits to people, mostly local communities, living in or near to the designated protected area.
- To facilitate recreation and appropriate small-scale tourism.

Distinguishing Features

- Category VI protected areas, uniquely amongst the IUCN categories system, have the sustainable use of natural resources as a *means* to achieve nature conservation, together and in synergy with other actions more common to the other categories, such as protection.
- Category VI protected areas aim to conserve ecosystems and habitats, together with associated cultural values and natural resource management systems. Therefore, this category of protected areas tends to be relatively large (although this is not obligatory).
- The category is not designed to accommodate large-scale industrial harvest.
- In general, IUCN recommends that a proportion of the area is retained in a natural condition, which in some cases might imply its definition as a no-take management zone. Some countries have set this as two-thirds; IUCN recommends that decisions need to be made at a national level and sometimes even at the level of individual protected areas.

Role in the Landscape/Seascape

- Category VI protected areas are particularly adapted to the application of landscape approaches.
- This is an appropriate category for large natural areas, such as tropical forests, deserts and other arid lands, complex wetland systems, coastal and high seas, boreal forests etc. – not only by establishing large protected areas, but also by linking with groups of protected areas, corridors or ecological networks.

- Category VI protected areas may also be particularly appropriate to the conservation of natural ecosystems when there are few or no areas without use or occupation and where those uses and occupations are mostly traditional and low-impact practices, which have not substantially affected the natural state of the ecosystem.

What makes Category VI Unique?

Allocation of category VI depends on long-term management objectives and also on local specific characteristics. The following table outlines some of the main reasons why category VI may be chosen in specific situations vis-à-vis other categories.

Category VI differs from the other categories in the following ways:	
Category Ia	Category VI protected areas do conserve biodiversity, particularly at ecosystem and landscape scale, but the aim would not be to protect them strictly from human interference. Although scientific research may be important, it would be considered a priority only when applied to sustainable uses of natural resources, either in order to improve them, or to understand how to minimize the risks to ecological sustainability.
Category Ib	Category VI protected areas in certain cases could be considered close to "wilderness", however they explicitly promote sustainable use, unlike the situation in category Ib wilderness areas where such use will be minimal and incidental to conservation aims. They also contribute to the maintenance of environmental services, but not only by exclusive nature conservation, as the sustainable use of natural resources can also contribute to the protection of ecosystems, large habitats, and ecological processes.
Category II	Category VI protected areas aim to conserve ecosystems, as complete and functional as possible, and their species and genetic diversity and associated environmental services, but differ from category II in the role they play in the promotion of sustainable use of natural resources. Tourism can be developed in category VI protected areas, but only as a very secondary activity or when they are part of the local communities' socio-economic strategies (e.g., in relation to ecotourism development).
Category III	Category VI protected areas might include the protection of specific natural or cultural features, including species and genetic diversity, among their objectives, whenever the sustainable use of natural resources is also part of the objectives, but they are more oriented to the protection of ecosystems, ecological processes, and maintenance of environmental services through nature protection and promotion of management approaches that lead to the sustainable use of natural resources.
Category IV	Category VI protected areas are more oriented to the protection of ecosystems, ecological processes, and maintenance of environmental services through nature protection and promotion of the sustainable use of natural resources. While category IV protected areas tend to prioritize active management, category VI promotes the sustainable use of natural resources.
Category V	Category V applies to areas where landscapes have been transformed as a result of long-term interactions with humans; category VI areas remain as predominantly natural ecosystems. The emphasis in category VI is therefore more on the protection of natural ecosystems and ecological processes, through nature protection and promotion of the sustainable use of natural resources.

Issues for Consideration

- Protection of natural ecosystems and promotion of sustainable use must be integrated and mutually beneficial; category VI can potentially demonstrate best management practices that can be more widely used.
- New skills and tools need to be developed by management authorities to address the new challenges that emerge from planning, monitoring and managing sustainable use areas.
- There is also need for development of appropriate forms of governance suitable for category VI protected areas and the multiple stakeholders that are often involved. Landscape-scale conservation inevitably includes a diverse stakeholder group, demanding careful institutional arrangements and approaches to innovative governance.

TRANSBOUNDARY PROTECTED AREA

A transboundary protected area (TBPA) is an ecological protected area that spans boundaries of more than one country or sub-national entity. Such areas are also known as transfrontier conservation areas (TFCAs) or peace parks.

Flags of Canada and the United States in Waterton-Glacier International Peace Park.

TBPAs exist in many forms around the world, and are established for various reasons. The preservation of traditional animal migration patterns, ensuring sufficient food and water sources for population growth, is a critical reason for the creation of TBPAs. However, TBPAs also encourage tourism, economic development and goodwill between neighbouring countries, as well as making it easier for indigenous inhabitants of the area to travel.

Types of Transboundary Protected Areas

TBPAs exist in various types of geographic configuration, with various levels of ecological protection, and with various levels of international cooperation. Additionally, different organizations employ different definitions for TBPAs. Julia Marton-Lefevre broadly defines TBPAs as "areas that involve a degree of cooperation across one or more boundaries between (or within) countries." The Southern Africa Development Community's Protocol on Wildlife Conservation and Law Enforcement defines Transfrontier Conservation Area as "the area or the component of a large ecological region that straddles the boundaries of two or more countries, encompassing one or more protected areas, as well as multiple resources use areas." The Global Transboundary Protected Areas Network (GTPAN) lists four types of "transboundary conservation areas:"

- Type 1: Transboundary Protected Area.
- Type 2: Transboundary Conservation Landscape and Seascape.
- Type 3: Transboundary Conservation Migration Area.
- Special designation: Park for Peace.

GTPAN defines a Transboundary Protected Area as "a clearly defined geographical space that includes protected areas that are ecologically connected across one or more international boundaries and involves some form of cooperation." GTPAN defines a Transboundary Conservation Landscape and Seascape as "an ecologically connected area that includes both protected areas and multiple resource use areas across one or more international boundaries and involves some form of cooperation."

GTPAN defines a Transboundary Conservation Migration Area as "wildlife habitats in two or more countries that are necessary to sustain populations of migratory species and involve some form of cooperation."

GTPAN defines a "Park for Peace" as "any of the three types of Transboundary Conservation Areas dedicated to the promotion, celebration and commemoration of peace and cooperation."

In many instances, individual TBPAs are part of broader international environmental or cultural programs. TBPAs can be World Heritage sites, Ramsar Wetlands, and UNESCO Biosphere Reserves.

Areas in the Conceptual Phase

The DMZ is a habitat for many threatened species on the Korean peninsula, including the black bear.

TBPA advocates by 2006 had identified additional sites for protection. Professor Saleem Ali of the University of Vermont noted that "numerous ecologically sensitive areas remain unprotected" and cites a 2006 "geographic information systems (GIS) study" that "found 104 transboundary wild areas involving 61 countries that are not formally part of any conservation park." Specific sites for proposed TBPAs include:

- Liuwa Plain: Kameia TFCA (Angola/Zambia).
- Niassa Reserve: Selous TFCA (Mozambique/Tanzania).
- Mnazi Bay: Quirimbas Transfrontier Marine Conservation Area (Mozambique/Tanzania).
- The demilitarised zone between North and South Korea is a potential TBPA/peace park. Professor Salem Ali notes that the DMZ has "become a sanctuary for wildlife, since there is no development activity occurring in the area." Nelson Mandela traveled to South Korea to promote such an idea. Ted Turner likewise traveled to the Korean peninsula to promote the idea in 2005.
- The Siachen Glacier region between India and Pakistan has been proposed as a peace park.
- A project for the Green Line Peace Park in Cyprus, between the Turkish Cypriot and Greek Cypriot territories, is being researched by Dr Anna Grichting at Harvard University.

As awareness of the importance of conserving the pristinity and ecology of Arctic region has increased, there has been a global call to declare the Arctic region as a global sanctuary/international peace park. The Save the Arctic campaign by Greenpeace, an environmental nonprofit organisation, has received online support from more than 5 million citizens from around the world.

Impacts on Peace and Conflict

Transboundary protected areas are also termed peace parks. They are supposed to facilitate cooperation and exchange between (adversary) countries, to improve livelihoods of local populations, to demonstrate the possibility of positive-sum interactions, and hence to support more peaceful international relations. There are several cases documented in which trans-boundary conservation contributed to conflict resolution (although it was not the main driver), such as in the Virunga region between the DR Congo, Rwanda and Uganda, around the Trifinio region between El Salvador and Honduras, and in the Cordillera del Cóndor region between Ecuador and Peru. According to a statistical analysis published in 2014, states that share a trans-boundary protected area are slightly less likely to engage with militarized disputes with each other. But the question remains whether the TBPA is a driver of consequence on better interstate relations in these cases. A more recent analysis triangulates data from various sources to show that international environmental cooperation (in the form of TBPAs and water treaties) increases the likelihood for reconciliation between states in conflict. The effect is, however, modest and contingent on a number of context factors such as high levels of environmental attention, internal political stability, a tradition of environmental cooperation and already ongoing processes of reconciliation.

However, a number of authors criticize that peace parks have a very limited impact on formal relations between states, but can accelerate conflicts on the local level, for instance by extending (authoritarian) state control, by prioritizing business and tourism over the interests of local populations, and by excluding local people from the protected areas. TBPAs can also stimulate (low-level) international conflicts, for instance about the sharing of revenues or the presence of human populations in the parks.

KEY BIODIVERSITY AREA

The Key Biodiversity Areas (KBA) approach helps to identify and designate areas of international importance in terms of biodiversity conservation using globally standardised criteria. KBAs extend the Important Bird Area (IBA) concept to other taxonomic groups and are now being identified in many parts of the world, by a range of organisations. Examples include Important Plant Areas (IPAs), Ecologically and Biologically Significant Areas (EBSAs) in the High Seas, Alliance for Zero Extinction (AZE) sites, Prime Butterfly Areas, Important Mammal Areas and Important Sites for Freshwater Biodiversity, with prototype criteria developed for freshwater molluscs and fish and for marine systems. The determination of KBAs often brings sites onto the conservation agenda that hadn't previously been identified as needing protection due to the nature of the two non-exclusive criteria used to determine them; vulnerability; and irreplaceability.

Objectives

- Develop technical and conservation capacity within individual countries and on a global scale.
- Develop partnerships between key organisations – both governmental and nongovernmental – concerned with site conservation.
- Build broad understanding of the process, and broad ownership of the final site list.
- Focus any new survey work on the most important gaps in knowledge.

INDIGENOUS AND COMMUNITY CONSERVED AREA

Indigenous and community conserved areas (ICCAs), or indigenous peoples' and community conserved territories and areas, are spaces de facto governed by indigenous peoples or local communities with evidently positive outcomes for the conservation of biological and cultural diversity. In ICCAs, the continuation, revival or modification of traditional practices (some of which are of ancient origin) and new initiatives succeed in protecting and restoring natural resources and cultural values in the face of new threats or opportunities. Some ICCAs are situated in remote ecosystems that have had minimum human influence, while others encompass areas of various regulations and magnitudes within regions strongly affected or modified by human occupation. ICCAs may or may not fit the IUCN definition of "protected area" but, when they do, they can fall into any IUCN protected area categories.

The following three characteristics are used to identify an ICCA:

- A strong relationship exists between an indigenous people or local community, and a specific site (territory, ecosystem, species habitat). This relationship is often embedded in the people's or community's sense of identity and dependence for livelihood and well-being.
- The indigenous people or local community is the major player in decision-making and implementation regarding the management of the site, implying that a local institution has the capacity to develop and enforce decisions (other stakeholders may collaborate as partners, especially when the land is owned by the state, but de facto decisions and management efforts are predominantly taken by the concerned people or community).
- The people's or community's management decisions and efforts lead to the conservation of habitats, species, genetic diversity, ecological functions/benefits

> and associated cultural values, even when the conscious objective of management is not conservation (i.e., it may be livelihoods, security, safeguarding cultural and spiritual values, etc).

The IUCN World Parks Congress of 2003 defined ICCAs as:

> "Natural and modified ecosystems containing significant biodiversity values and ecological services, voluntarily conserved by (sedentary and mobile) indigenous and local communities, through customary laws or other effective means."

This definition is recognised by the United Nations Environment Programme World Conservation Monitoring Centre (UNEP-WCMC) and is elaborated on in the UNEP-WCMC ICCA Registry Handbook as a type of Protected Area (including Marine Protected Areas) in which native peoples initiate the creation and are owners and managers.

Dynamics

A village association runs an ICCA-based sustainable use enterprise in Costa Rica.

A crucial feature of ICCAs is their diversity. The conservation practices of indigenous peoples and local communities depend on an astonishing variety of meanings and values underpinned by the relationship between humans and the natural environment, and find their expression in different ICCAs all over the world. While all ICCAs by definition embody precious bio-cultural diversity in a voluntary and self-organised way, the related beliefs, practices, and institutions are all context-specific. Moreover, as lively socio-cultural phenomena, ICCAs change in correspondence with history and society. Some disappear, others survive in old or new forms, and some will emerge anew. Most systems by which contemporary indigenous peoples and local communities govern and manage their natural resources are a blending of old and new knowledge, practices, tools and values of different origin.

In the struggle to cope with the scale and pace of socio-cultural change, some ICCA institutions have been replaced by state governance or are under threat of being so. Similarly in some cases, change has been powerful enough to affect the community's capacity to manage the local resources in a sustainable way and genuine local ICCAs are just a memory or very much struggling to remain effective. Yet in others, even powerful change has been unable to destroy them: More complex ICCAs, capable of taking advantage of new conditions and establishing new alliances have emerged from the pre-existing ones.

Over the last two centuries, the formal policies and practices that dominate conservation and development have largely ignored ICCAs or actively threaten them. Even today, while neglect and harm give way to emerging recognition and support, the interface between state-based institutions and the customary institutions of indigenous peoples and local communities remains riddled with conflicts. Some relationships are respectful, but many are affected by misunderstandings, mistrust which may threaten the success of well-intentioned initiatives. In fact, despite the current serious interest on individual ICCAs and community conservation in general, two opposing stereotypes continue to plague conservation: The romantic view of indigenous peoples and traditional communities Soing in total harmony with nature; and the view of people as "parasites", necessarily degrading the ecosystems in which they live.

Threats

Despite increasing recognition of ICCAs in international conservation policies, there is still great neglect in terms of their effective and appropriate recognition in national policies and practices. When they have no legal recognition within a country, they may also not be recognised or respected by private entities and neighbouring communities. In such cases, ICCAs are vulnerable through land and water being appropriated or "re-allocated" for an alternative use. To non-members of the relevant communities, many ICCAs appear as natural, "unmanaged" and "unutilised" ecosystems — all the more coveted for resource extraction. ICCAs may also suffer as a result of changing value systems, increased pressure on natural resources and other internal tensions In general, ICCAs are exposed to both external and internal threats.

External

- Imposed development and resource exploitation processes, such as mining and fossil fuel extraction (particularly important as even if indigenous peoples and local communities possess land rights, government usually reserve for themselves the use of sub-soil resources), logging, tree plantation, industrial fishing, sea dredging, land conversion to large-scale grazing or agriculture (including agrofuel plantations), water diversions and drainage works, urbanisation and major infrastructure (roads, ports, airports, tourism).
- Expropriation of community land through nationalisation, privatisation, and conservation initiatives, particularly the creation of state-governed protected areas.

- War, violent conflicts or movements of refugees.
- Territorial encroachment by or conflicts with other communities and municipalities.
- Inappropriate forms of recognition, such as that which imposes top-down institutional arrangements and thereby devalues and de-motivates traditional governance institutions.
- Imposition of unaffordable taxes and other fiscal burdens.
- Active acculturation of communities i.e.education programmes disrespectful of local cultures, livelihoods and values, or evangelisation programmes of different faiths.
- Divisions and conflicts fuelled by party politics (often actively promoted from outside) or by sudden influx of funds strengthening or creating local inequities.
- Poaching and unauthorised extraction of timber and plant resources.
- Air and water pollution through discharge of waste residuals (e.g. via acid rain, chemical pollution from upstream mining or run-off of chemical inputs from agriculture) and the spread of invasive or exotic species.
- Extreme natural events and catastrophes, including droughts, floods, forest fires, hurricanes, earthquakes and tsunamis, some of which are related to human transformation of the landscape, waterways and climate.

Internal

- Changing values, acculturation and integration into dominant society, leading to commoditisation of nature and culture and, ultimately, the loss of traditional knowledge, locally adapted management practices and governance institutions — all with particular impact on the younger generations.
- Increasing pressure on resources — in particular those that lead to the substitution of local subsistence and solidarity economies with the market economy.
- Persistent or new inequalities between economic and social classes and gender groups within the community, leading to conflicts about management of natural resources and elite capture of conservation benefits.
- Depopulation through migration because of new economic opportunities, social conflicts and political pressures.
- Progressive loss of food sovereignty and traditional medical systems, weakening traditional communities because of poorer health and nutrition.

In practice, threats cannot necessarily be segregated into "external" and "internal" categories, as community members may be active participants in external processes, and exogenous forces may drive internal processes. For instance, a main driver of change that powerfully combines external and internal threats are new opportunities to access and use natural resources for profit-making activities. These may bring in welcome funds for a variety of development needs but can also be a door for corruption and mis-governance, ushering divisions, conflicts and social disruption. As the disparity of power in modern societies increases exponentially, many indigenous peoples and local communities, at the bottom of the ladder, have fewer and fewer chances to resist. In some countries they are even denied legal existence as "peoples" and "communities", and denied the chance of owning or possessing land rights and natural resources collectively, one of the last barriers to individual weakness and greed. In theory, the recognition of the many values of ICCAs will help in the broader struggles for human rights and indigenous peoples' rights, and contribute to foster more equitable and sustainable societies.

Legislation

Discussing ICCA governance issues in Coron Island – one of the first ICCAs to be fully recognized as Ancestral Domain (Terrestrial and marine area), under the collective governance of its indigenous peoples (Tagbanwa) in Palawan.

It is said that ICCAs cover as much land as government protected areas (as well as existing for much longer), yet tens of thousands of sites are not yet recognised by governments and still face neglect within official conservation systems and government policies and legislation. A lack of political and legal support often hampers community efforts at maintaining ICCAs through traditional means.

While the "legitimacy" of ICCAs is rooted in the values and meanings they possess for the most directly concerned peoples and communities, their "legality" and their broad recognition and support by society at large are rooted in a process that takes strength from international conventions and agreements. This process originated relatively recently. At the Vth World Parks Congress conservation professionals systematized for the first time the concept of "governance of protected areas" and clarified that indigenous peoples and local communities – a crucial actor in conservation – should be fully recognized in their governance role. At the same Congress a breakthrough was made by indigenous peoples – and mobile indigenous peoples in particular – effectively arguing that the respect of their rights would actually advance, rather than diminish, conservation outcomes. Shortly after the Durban Congress, the Convention on Biological Diversity, at its COP 7 meeting in Kuala Lumpur, approved the CBD Programme of Work on Protected Areas (PoWPA). PoWPA supports a "new approach" to protected areas, calling for attention to governance types and quality, equity in conservation, and indigenous peoples' rights.

Numerous IUCN Resolutions attest to the will of IUCN members to recognize and support ICCAs were approved at the Barcelona World Conservation Congress of 2008, and IUCN publications were developed to back this up technically. WCC4 in Barcelona also approved new IUCN technical guidelines for protected areas, explicitly stating that different governance types – including ICCAs – can fully contribute to developing national protected area systems.

CBD COP 8 and CBD COP 9 reviewed the implementation of PoWPA and stressed the need to engage more forcefully in the element dedicated to 'Governance, Participation, Equity and Benefit Sharing'. This was also reflected in the statement of recommendations that the May 2010 meeting of CBD SBSTTA in Nairobi submitted to COP 10. Noticeably, CBD SBSTTA delegates made specific recommendations about ICCA recognition, clarifying, for instance, that "mechanisms for recognition should respect the customary governance systems that have maintained ICCAs over time".

At CBD COP 10 in Nagoya, decision X/31 stressed again the role of indigenous (peoples) and community conserved (territories and) areas and invited Parties to recognise their organisations and contributions. COP 10 also agreed on a number of biodiversity targets for 2020. CBD Aichi Target 11 foresees that: "By 2020 at least 17 per cent of terrestrial and inland water, and 10 per cent of coastal and marine areas, especially areas of particular importance for biodiversity and ecosystem services, are conserved through effectively and equitably managed, ecologically representative and well-connected systems of protected areas and other effective area-based conservation measures, and integrated into the wider landscapes and seascapes." ICCAs clearly have a role to fulfil Aichi Target 11 as both protected areas and as "other effective area-based conservation measures".

BIOSPHERE RESERVES

Biosphere reserves are areas comprising terrestrial, marine and coastal ecosystems. Each reserve promotes solutions reconciling the conservation of biodiversity with its sustainable use.

Biosphere reserves are 'Science for Sustainability support sites' – special places for testing interdisciplinary approaches to understanding and managing changes and interactions between social and ecological systems, including conflict prevention and management of biodiversity. Biosphere reserves are nominated by national governments and remain under the sovereign jurisdiction of the states where they are located. Their status is internationally recognized.

There are 651 biosphere reserves in 120 countries, including 15 transboundary sites. They are distributed as follows:

- 67 in 28 countries in Africa.
- 28 in 11 countries in the Arab States.
- 137 in 24 countries in Asia and the Pacific.
- 297 in 36 countries in Europe and North America.
- 122 in 21 countries in Latin America and the Caribbean.

Zones in Biosphere Reserve

Biosphere reserves have three interrelated zones that aim to fulfil three complementary and mutually reinforcing functions:

- The core area s comprises a strictly protected ecosystem that contributes to the conservation of landscapes, ecosystems, species and genetic variation.
- The buffer zone surrounds or adjoins the core areas, and is used for activities compatible with sound ecological practices that can reinforce scientific research, monitoring, training and education.
- The transition area is the part of the reserve where the greatest activity is allowed, fostering economic and human development that is socio-culturally and ecologically sustainable.

Main Characteristics of Biosphere Reserves

- Achieving the three international functions: Conservation, development and logistic support.
- Outpacing traditional confined conservation zones, through appropriate zoning schemes combining core protected areas with zones where sustainable development is fostered by local dwellers and enterprises with often highly innovative and participative governance systems.
- Focusing on a multi-stakeholder approach with particular emphasis on the involvement of local communities in management.

- Fostering dialogue for conflict resolution of natural resource use.
- Integrating cultural and biological diversity, especially the role of traditional knowledge in ecosystem management.
- Demonstrating sound sustainable development practices and policies based on research and monitoring.
- Acting as sites of excellence for education and training.
- Participating in the World Network.

References

- Category-ia-strict-nature-reserve, protected-areas-about-protected-areas-categories: iucn.org, Retrieved 09 May, 2020
- Rollins, John (2004). Caves Of The Canadian Rockies And Columbia Mountains. Surrey, Canada: Rocky Mountain Books. ISBN 978-0-92110-294-6. Retrieved 25 May, 2015
- Lahren, Larry (2006). Homeland: An archaeologist's view of Yellowstone Country's past. Cayuse Press. p. 161. ISBN 978-0-9789251-0-9
- Bonnett, A. (2016). The Geography of Nostalgia: Global and Local Perspectives on Modernity and Loss. Routledge. p. 68. ISBN 978-1-315-88297-0
- What-is-biosphere-reserve-complete-detail-definition-of-biosphere-reserve: natureconservation.in, Retrieved 10 January, 2020

3

Marine Protected Areas

Marine protected areas include the protected areas of seas, oceans, estuaries, etc. These protected areas restrict human activities for protecting natural resources and conserving marine biodiversity. Marine reserves, marine parks, shark and whale sanctuaries, etc. are such protected areas. This chapter has been carefully written to provide an easy understanding of marine protected areas.

Marine protected areas (MPAs) play an important role in protecting and conserving our global ocean ecosystems. The history and development of MPAs around the world are addressed, from early stages to current levels of protection today, through to globally recognized progressive targets for ocean conservation. This outlines the positive benefits of MPAs, their different forms and types, and how these are categorized by the International Union for Conservation of Nature (IUCN). Recent trends involve protection of high sea areas beyond national jurisdiction, the World Heritage listing of sites with Outstanding Universal Value, and a growing recognition of the importance of protecting large-scale marine areas > 150,000 km^2.

Marine protected areas (MPAs) are a regulatory tool for conserving the natural or cultural resources of the ocean and for managing human uses through zoning. MPAs may also be referred to as marine parks, sanctuaries, reserves, or closures; the latter two terms are used most commonly in the context of fisheries management.

At a conceptual level, zoning in the ocean involves the spatial segregation of a marine area in which certain uses are regulated or prohibited. This general definition might apply to any marine area in which a set of human uses are given preference over others. For example, by law the US President may set aside hydrocarbon deposits on the US outer Continental Shelf as 'petroleum reserves.' However, the typical use of the term 'protected' implies that a primary focus of an MPA is on the conservation of either individual species and their habitats or ecological systems and functions through the regulation of 'extractive' or potentially polluting commercial uses, such as fishery harvests, waste disposal, and mineral development, among others.

MPAs are frequently considered to be a fishery management measure, but they may be used for other purposes as well. For instance, in 1975, the first US national marine sanctuary was created around the wreck of the U.S.S. Monitor, a civil war vessel, located off the coast of North Carolina. The sanctuary was established to prevent commercial 'treasure' salvage and looting of the shipwreck, to regulate recreational diving, and to promote archaeological studies.

Size

Although there is no discernible size limitation, the issue of geographic scale may be another defining characteristic of MPAs. On the tidelands of US coastal states, for example, the 'public trust doctrine' gives preference in the common law to transitory public uses, typically navigation, fishing, and hunting, over permanent private uses, such as constructing a dock. Yet the tidelands, which are quite extensive, are not referred to as an MPA. Some fishery closures can be quite large, and we would classify these as one type of MPA. The Great Barrier Reef Marine Park in Australia is the largest MPA in the world, measuring 344 million km². Most of the world's existing MPAs are much smaller, however, and focused on unique ocean features or sites, such as coral reefs or underwater banks. The World Bank estimates the median size of a sample of about one thousand of the world's MPAs to be 15 840 km².

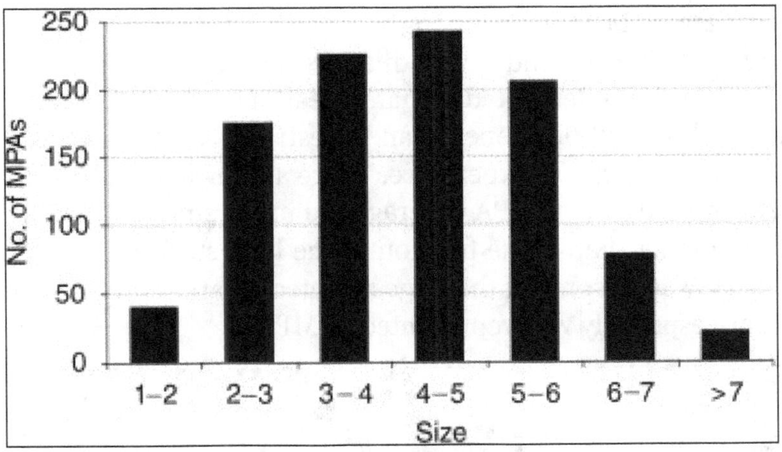

Worldwide size distribution of marine protected areas (n=991). Sizes are grouped by km² to the powers of ten.

Number

Worldwide, MPAs have become a popular form of ocean management, and their use has expanded exponentially since they were first introduced in the late nineteenth century. The trend in the establishment of MPAs follows on the heels of a more general trend in the regulation of ocean uses, as an MPA represents merely a form of governance distinguishable geographically by type or severity of regulation. Regulation of the ocean has become necessary as human uses of the ocean have increased in scale and variety and as conflicts among mutually exclusive uses and users have arisen.

With the expansion in the establishment of MPAs, marine scientists and policy experts have begun to take a closer look at the likely benefits and opportunity costs associated with zoning the ocean, and several recent studies have emerged. In particular, as ecological models of the marine environment become more realistic, marine policy analysts can begin to make more sophisticated examinations of how to choose among competing human uses, given the constraints presented by the natural system.

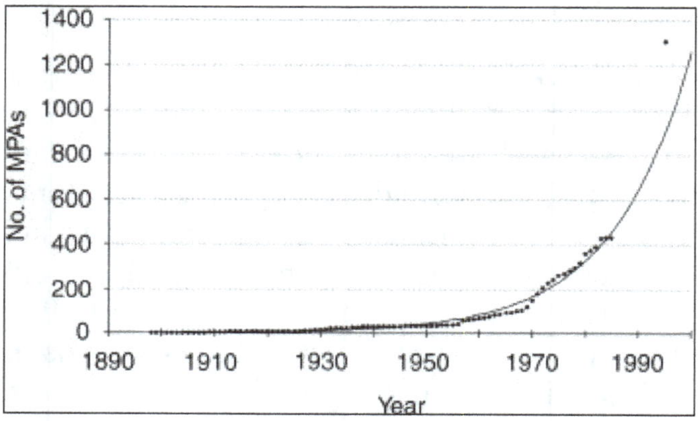

Cumulative worldwide growth in the number of marine protected areas and estimated logarithmic trend.

Marine protected areas (MPAs) are among the first to apply these principles due to the immediate nature of changes under way due tocoral bleaching. MPAs serve multiple purposes and have been implemented in many areas of the planet. However, in general, MPA networks are less well-developed than terrestrial protected areas, both in extent of coverage and in management experience. There are far fewer MPAs than terrestrial protected areas, and the area of MPA coverage is a much smaller fraction of the overall ocean area of the planet than is the fraction of the land surface covered by terrestrial protected areas. However, some of the largest protected areas on the planet are MPAs, and coral reefs are especially well represented in MPAs.

Healthy and Bleached Coral Reefs.

MPA response to coral bleaching is built around the principles of resistance and resilience. These principles are applied on different spatial scales to help guide MPA system planning and management planning of individual MPAs, respectively. In this case, resistance refers to reefs that are resistant to bleaching, while resilience refers to reefs that are more likely to recover once damaged.

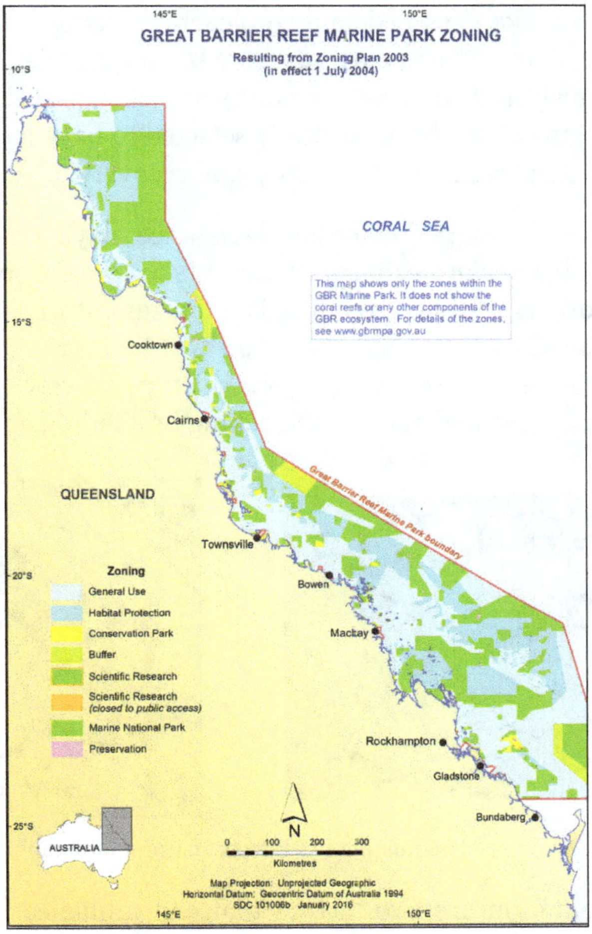

Zoning Map for the Great Barrier Reef Marine Park.

The Great Barrier Reef Marine Park is a marine protected area that has experienced extensive coral bleaching. In response to bleaching events and other management issues, a zoning plan for the park reflects permitted uses that best integrate climate change with other park management objectives. Tourism is excluded in some areas to facilitate postbleaching recovery.

At the broad-scale, system planning level, bleaching resistance and bleaching resilience are used to help prioritize sites for protection. These are not the only, nor even the main, prioritization factors, but between otherwise equal priority sites, resistant and resilient sites are selected over low-resistance or low-resilience sites. Although biological value and irreplaceability are primary criteria, unless a site harbors completely irreplaceable attributes (e.g., species found only at that site and nowhere else in the world), there is little point in protecting a reef that will be destroyed by bleaching. Hence, resistant and resilient sites get preference.

Sites may be more or less resistant and resilient to coral bleaching based on factors such as currents, upwelling, and previous exposure of reefs to bleaching. Cool currents or upwellings tend to make sites resistant to bleaching because they maintain cooler surface water temperatures. Warm currents have the opposite effect. Corals that have been bleached previously and survived (mortality in bleached corals ranges from 60 to 90% in most regions) may have been naturally selected to have more resistant zooxanthelae or other natural mechanisms of survival and recovery.

At the site management level, resilience and resistance may be applied in zoning and management decisions. Local upwellings or currents may make some parts of an MPA more resilient than others. Physical shading is important at this scale as well. For instance, reefs in the shadow of a large mountain will be more resistant than reefs in continual exposure to the sun. Sedimentation, which negatively affects reefs in heavy doses, may actually increase resistance to bleaching by shading and cooling reefs.

Shading and Sedimentation Effects on Coral Bleaching.

Shading and Sedimentation are two factors that can influence the severity of coral bleaching. Sedimentation stresses corals and may exacerbate bleaching effects, whereas shading protects corals from synergies of high temperatures and photic effects, thereby reducing the probability of bleaching.

Once reefs do bleach, resilience is an important factor in reserve management. Reefs that are more likely to recover become a source of recolonization of corals for areas in which mortality is high. Zooxanthelae from resilient reefs may eventually colonize damaged reefs and increase their resilience. Prioritizing these areas for protection is therefore as high a priority at the local scale as it is at the regional (system planning/site selection) scale.

Management of MPAs can reinforce resilience and recovery. When an area bleaches, temporary tourism closures can help the reef recover. Where recreational fisheries or subsistence harvest is allowed, temporary closures may also help ease pressure on coral-dependent species whose populations crash following bleaching, allowing them to recover once corals re-establish. Lack of protective management can result in the exacerbation of the damage of bleaching, thus causing a downward spiral in reef condition that may ultimately result in the replacement of the corals in the system with algae.

Marine Protected Areas.

Healthy marine systems such as these can be one of the major benefits of marine protected areas (MPAs). MPAs can improve food web health and reduce chances of coral bleaching by decreasing synergistic pressures such as fishing and tourism overuse.

Nontropical MPAs must also carefully consider climate change in management. Extensive evidence from cold-water fisheries and past climate change indicates that pelagic fisheries and benthic communities experience major shifts linked to climate. Recent changes due to loss of sea ice in the North Bering Sea provide a compelling example of the need for additional MPAs.

Sea ice retreat in warming sea surface temperatures is driving major ecological changes in the North Bering Sea near St. Lawrence Island. The disappearance of sea ice in warmer water temperatures makes a more active ecosystem in the water column in this region. Plankton that used to fall through the water column and be deposited in the benthos now are caught up in the food web of the water column, greatly reducing the energy and nutrients reaching benthic communities. As a result, there has been a large-scale shift in the species composition of benthic communities, with mollusks decreasing in dominance and nutritional quality and being replaced by brittle stars.

Walrus and spectacled eider that feed on these benthic communities are therefore faced with greatly declining nutrition. At the same time, the retreat of sea ice means that the species spend more time resting in water than on ice. Thermal losses in water are much greater than those in air, so the switch from ice resting to water resting means that eider and walrus are burning much more energy at rest at the same time that they are getting less bivalve food from the bottom. If the species try to follow sea ice to maintain their thermal balance, they find themselves in much deeper waters farther from the continental shelf, which means they must dive deeper to feed.

Walrus and eider populations have declined dramatically due to these deteriorating conditions. Spectacled eider populations have declined by up to 96% in areas of Alaska and Canada. Young walrus are being separated from their mothers as sea ice literally melts out from under them. Pups drift away while their mothers are on deep feeding dives. Lost walrus pups have been observed by research vessels in open ocean thousands of miles from normal walrus breeding grounds.

To further complicate the situation, there is a sill of cold water behind a ridge on the ocean floor off St. Lawrence Island that excludes groundfish from the North Bering Sea. This cold-water sill is breaking down as water temperatures warm. As it disappears, groundfish will enter the North Bering Sea and compete with walrus and eider for mollusks.

Bottom trawling fisheries will likely follow groundfish when they enter the North Bering Sea, disrupting the bottom and further reducing mollusk food sources. This human disturbance may be the coup de grace for declining walrus and eider populations if fisheries are allowed to enter the area unchecked.

An MPA in these waters is needed to protect the bottom habitat for mollusk populations to maintain food sources for walrus and eiders. This protection cannot reverse the loss of sea ice, the warming of North Bering Sea waters, or the breakdown of the cold-water sill. An MPA can, however, prevent the decline of bottom habitat due to trawling and so minimize further losses of diving species.

Marine Conservation

Coral reefs have a great amount of biodiversity.

Marine conservation, also known as ocean conservation, refers to the study of marine plants and animal resources and ecosystem functions. It is the protection and preservation of ecosystems in oceans and seas through planned management in order to prevent the exploitation of these resources. Marine conservation is driven by the manifested negative effects being seen in our environment such as species loss, habitat degradation and changes in ecosystem functions and focuses on limiting human-caused damage to marine ecosystems, restoring damaged marine ecosystems, and preserving vulnerable species and ecosystems of the marine life. Marine conservation is a relatively new discipline which has developed as a response to biological issues such as extinction and marine habitats change.

Marine conservationists rely on a combination of scientific principles derived from marine biology, oceanography, and fisheries science, as well as on human factors such as, demand for marine resources and marine law, economics and policy, in order to determine how to best protect and conserve marine species and ecosystems. Marine conservation may be described as a sub-discipline of conservation biology.

Coral Reefs

Coral reefs are the epicenter of immense amounts of biodiversity and are a key player in the survival of entire ecosystems. They provide various marine animals with food, protection, and shelter which keep generations of species alive. Furthermore, coral reefs are an integral part of sustaining human life through serving as a food source (i.e., fish and mollusks) as well as a marine space for ecotourism which provides economic benefits. Also, humans are now conducting research regarding the use of corals as new potential sources for pharmaceuticals (i.e. steroids and anti-inflammatory drugs).

Unfortunately, because of the human impact on coral reefs, these ecosystems are becoming increasingly degraded and in need of conservation. The biggest threats include overfishing, destructive fishing practices, sedimentation, and pollution from land-based sources. This, in conjunction with increased carbon in oceans, coral bleaching, and diseases, means that there are no pristine reefs anywhere in the world. Up to 88% of coral reefs in Southeast Asia are now threatened, with 50% of those reefs at either "high" or "very high" risk of disappearing, which directly affects the biodiversity and survival of species dependent on coral.

This is especially harmful to island nations such as Samoa, Indonesia, and the Philippines, because many people there depend on the coral reef ecosystems to feed their families and to make a living. However, many fishermen are unable to catch as many fish as they used to, so they are increasingly using cyanide and dynamite in fishing, which further degrades the coral reef ecosystem. This perpetuation of bad habits simply leads to the further decline of coral reefs and therefore perpetuates the problem. One way of stopping this cycle is by educating the local community about why the conservation of marine spaces that include coral reefs is important.

Human Impact

Increasing human populations have resulted in increased human impact on ecosystems. Human activities has resulted in an increased extinction rate of species which has caused a major decrease in biological diversity of plants and animals in our environment. These impacts include increased pressure from fisheries including reef degradation and overfishing as well as pressure from the tourism industry which has increased over the past few years. The deterioration of coral reefs is mainly linked to human activities – 88% of reefs are threatened, including excessive amounts of CO_2 (carbon dioxide) emissions. Oceans absorb approximately 1/3 of the CO_2 produced by humans, which has detrimental effects on the marine environment. The increasing levels of CO_2 in oceans change the seawater chemistry by decreasing the pH, which is known as ocean acidification.

Oil spills also impact marine environments, contributing to marine pollution as a result of human activity. The effects of oil on marine fish have been studied following major spills in the United States.

Shipping is a major vector for the introduction of exotic marine species, some of which can become overabundant and transform ecosystems. Collisions with ships can also be fatal for whales and can impact on the viability of whole populations, including the right whale population off the east coast of the United States.

Overabundance

Overabundance can occur when the population of a certain species cannot be controlled naturally or by human intervention. The domination of one species can create an imbalance in an ecosystem, which can lead to the demise of other species and of the habitat. Overabundance occurs predominately among invasive species.

Introduced Species

The international shipping trade has led to the establishment of many marine species beyond their native ranges. Some of these can have adverse consequences, such as the North pacific seastar which was introduced to Tasmania, Australia. Vectors for the translocation of organisms include hull biofouling, the dumping of ballast water and dumping of water from marine aquaria. A tank of ballast water is estimated to contain around 3,000 non-native species. Once established, it is difficult to eradicate an exotic organism from an ecosystem.

The San Francisco Bay is one of the places in the world that is the most impacted by foreign and invasive species. According to the Baykeeper organization, 97 percent of the organisms in the San Francisco Bay have been compromised by the 240 invasive species that have been brought into the ecosystem. Invasive species in the bay such as the Asian clam have changed the food web of the ecosystem by depleting populations of native species such as plankton. The Asian clam clogs pipes and obstructs the flow of water in electrical generating facilities. Their presence in the San Francisco Bay has cost the United States an estimated one billion dollars in damages.

Techniques

Strategies and techniques for marine conservation tend to combine theoretical disciplines, such as population biology, with practical conservation strategies, such as setting up protected areas, as with marine protected areas (MPAs) or Voluntary Marine Conservation Areas. These protected areas may be established for a variety of reasons and aim to limit the impact of human activity. These protected areas operate differently which includes areas that have seasonal closures and permanent closures as well as multiple levels of zoning that allow people to carryout different activities in separate areas; including, speed, no take and multi-use zones.

Other techniques include developing sustainable fisheries and restoring the populations of endangered species through artificial means.

Another focus of conservationists is on curtailing human activities that are detrimental to either marine ecosystems or species through policy, techniques such as fishing quotas, like those set up by the Northwest Atlantic Fisheries Organization, or laws such as those listed. Recognizing the economics involved in human use of marine ecosystems is key, as is education of the public about conservation issues. This includes educating tourists that come to an area who might not be familiar with certain regulations regarding the marine habitat. One example of this is a project called Green Fins based in Southeast Asia that uses the scuba diving industry to educate the public. This project, implemented by UNEP, encourages scuba diving operators to educate their students about the importance of marine conservation and encourage them to dive in an environmentally friendly manner that does not damage coral reefs or associated marine ecosystems.

Technology and Halfway Technology

Marine conservation technologies are used to protect endangered and threatened marine organisms and habitat. These technologies are innovative and revolutionary because they reduce by-catch, increase the survivorship and health of marine life and habitat, and benefit fishermen who depend on the resources for profit. Examples of technologies include marine protected areas (MPAs), turtle excluder devices (TEDs), autonomous recording unit, pop-up satellite archival tag, and radio-frequency identification (RFID). Commercial practicality plays an important role in the success of marine conservation because it is necessary to cater to the needs of fishermen while also protecting marine life.

Pop-up satellite archival tag (PSAT or PAT) plays a vital role in marine conservation by providing marine biologists with an opportunity to study animals in their natural environments. These are used to track movements of (usually large, migratory) marine animals. A PSAT is an archival tag (or data logger) that is equipped with a means to transmit the collected data via satellite. Though the data are physically stored on the tag, its major advantage is that it does not have to be physically retrieved like an archival tag for the data to be available, making it a viable independent tool for animal behavior studies. These tags have been used to track movements of ocean sunfish, marlin, blue sharks, bluefin tuna, swordfish and sea turtles. Location, depth, temperature, and body movement data are used to answer questions about migratory patterns, seasonal feeding movements, daily habits, and survival after catch and release.

Turtle excluder devices (TEDs) remove a major threat to turtles in their marine environment. Many sea turtles are accidentally captured, injured or killed by fishing. In response to this threat the National Oceanic and Atmospheric Administration (NOAA) worked with the shrimp trawling industry to create the TEDs. By working with the industry they insured the commercial viability of the devices. A TED is a series of bars that is placed at the top or bottom of a trawl net, fitting the bars into the "neck" of the shrimp trawl and acting as a filter to ensure that only small animals may pass through. The shrimp will be caught but larger animals such as marine turtles that become caught by the trawler will be rejected by the filter function of the bars.

Similarly, halfway technologies work to increase the population of marine organisms. However, they do so without behavioral changes, and address the symptoms but not the cause of the declines. Examples of halfway technologies include hatcheries and fish ladders.

Laws and Treaties

International laws and treaties related to marine conservation include the 1966 Convention on Fishing and Conservation of Living Resources of the High Seas. United States laws related to marine conservation include the 1972 Marine Mammal Protection Act, as well as the 1972 Marine Protection, Research and Sanctuaries Act, which established the National Marine Sanctuaries program.

In 2010, the Scottish Parliament enacted new legislation for the protection of marine life with the Marine (Scotland) Act 2010. Its provisions include marine planning, marine licensing, marine conservation, seal conservation, and enforcement. Since 2006, united nations introduce vulnerable marine ecosystem concept for the management of deep-sea fisheries in the areas beyond national jurisdiction. This concept has been transposed by the European parliament for Atlantic European waters.

Extinct and Endangered Species

Marine Mammals

Baleen whales were predominantly hunted from 1600 through the mid-1900s, and were nearing extinction when a global ban on commercial whaling was put into effect in 1986 by the IWC (International Whaling Convention). The Atlantic gray whale, last sighted in 1740, is now extinct due to European and Native American Whaling. Since the 1960s the global population of monk seals has been rapidly declining. The Hawaiian and Mediterranean monk seals are considered to be one of the most endangered marine mammals on the planet, according to the NOAA. The last sighting of the Caribbean monk seal was in 1952, and it has now been confirmed extinct by the NOAA. The vaquita porpoise, discovered in 1958, has become the most endangered marine species. Over half the population has disappeared since 2012, leaving 100 left in 2014. The vaquita frequently drowns in fishing nets, which are used illegally in marine protected areas off the Gulf of Mexico.

Sea Turtles

In 2004, the Marine Turtle Specialist Group (MTSG), from the International Union for Conservation of Nature (IUCN), ran an assessment which determined that green turtles were globally endangered. Population decline in ocean basins is indicated through data collected by the MTSG that analyzes abundance and historical information on the species. This data examined the global population of green turtles at 32 nesting sites, and determined that over the last 100–150 years there has been a 48–65 percent decrease in the number of mature nesting females. The Kemp's ridley sea turtle population fell in 1947 when 33,000 nests, which accounted for 80 percent of the population, were collected and sold by villagers in Racho Nuevo, Mexico. In the early 1960s only 5,000 individuals were left, and between 1978 and 1991, 200 Kemp's Ridley Turtles nested annually. In 2015, the World Wildlife Fund and National Geographic Magazine named the Kemp's ridley the most endangered sea turtle in the world, with 1000 females nesting annually.

Fish

In 2014, the IUCN moved the Pacific bluefin tuna from "least concerned" to "vulnerable" on a scale that represents level of extinction risk. The Pacific bluefin tuna is targeted by the fishing industry mainly for its use in sushi. A stock assessment released in 2013 by the International Scientific Committee for Tuna and Tuna-Like Species in the North Pacific Ocean (ISC) shows that the Pacific bluefin tuna population dropped by 96 percent in the Pacific Ocean. According to the ISC assessment, 90 percent of the Pacific bluefin tuna caught are juveniles that have not reproduced.

Between 2011 and 2014, the European eel, Japanese eel, and American eel were put on the IUCN red list of endangered species. In 2015, the Environmental Agency concluded that the number of European eels has declined by 95 percent since 1990. An Environmental Agency officer, Andy Don, who has been researching eels for the past 20 years, said, "There is no doubt that there is a crisis. People have been reporting catching a kilo of glass eels this year when they would expect to catch 40 kilos. We have got to do something."

Marine Plants

Johnson's seagrass, a food source for the endangered green sea turtle, is the scarcest species in its genus. It reproduces asexually, which limits its ability to populate and colonize habitats. Data on this species is limited, but it is known that since the 1970s there has been a 50 percent decrease in abundance.

MARINE PROTECTED AREA NETWORK

A Marine Protected Area Network or MPA network is a network of Marine Protected Areas or Marine Reserves.

A Marine Protected Area Network can be defined as "a collection of individual MPAs or reserves operating cooperatively and synergistically, at various spatial scales, and with a range of protection levels that are designed to meet objectives that a single reserve cannot achieve". Such a network can include several MPAs of different sizes, located in critical habitats, containing components of a particular habitat type or portions of different kinds of important habitats, and interconnected by the movement of animals and plant propagules.

A Marine Protected Area Network is usually established to improve fish catch, to conserve biodiversity, or for a combination of these two reasons. They are usually placed so that larvae can migrate from MPAs to other, more impacted areas. It provides a framework that unifies the central aims of conservation and fishery management, while also meeting other human needs such as maintenance of coastal water quality, shoreline protection, education, research and recreational opportunities.

MPA networks have been defined as "A group of MPAs that interact with one another ecologically and socially form a network".

These networks are intended to connect individuals and MPAs and promote education and cooperation among various administrations and user groups. "MPA networks are, from the perspective of resource users, intended to address both environmental and socio-economic needs, complementary ecological and social goals and designs need greater research and policy support".

Filipino communities connect with one another to share information about MPAs, creating a larger network through the social communities' support. Emerging or established MPA networks can be found in Australia, Belize, the Red Sea, Gulf of Aden and Mexico.

In Scotland, the Scottish Marine Protected Area Network covers approximately 22 % of Scotland's seas. It consists of 231 sites protected by a variety of different conservation designations, many of which are the same as those used on land, such as Special Protection Areas (SPA) and Sites of Special Scientific Interest (SSSI). In addition to the statutory MPAs, five further sites are recognised as forming part of the Scottish MPA network, being categorised as "other area based measures": such areas, although not specifically created for nature conservation purposes, are considered to contribute to the protection of marine biodiversity.

MARINE RESERVE

A marine reserve is a type of marine protected area that has legal protection against fishing or development. As of 2007 less than 1% of the world's oceans had been set aside in marine reserves. Benefits include increases in the diversity, density, biomass, body size and reproductive potential of fishery and other species within their boundaries.

Cape Rodney-Okakari Point, Goat Island Marine Reserve.

As of 2010, scientists had studied more than 150 marine reserves in at least 61 countries and monitored biological changes inside the reserves. The number of species in each study ranged from 1 to 250 and the reserves ranged in size from 0.006 to 800 square kilometers (0.002 to 310 square miles). In 2014, the World Parks Association adopted a target of establishing no-take zones for 30% of each habitat globally.

Design

A review of studies of 34 families (210 species) of coral reef fishes demonstrates that the design of a marine reserve has important implications for its ability to protect habitat and focal species.

Size and Shape

Effective reserves included habitats that support the life history of focal species (e.g.

home ranges, nursery grounds, migration corridors and spawning aggregations), and were located to accommodate movement patterns among them.

Movement patterns (home ranges, ontogenetic shifts and spawning migrations) vary among and within species, and are influenced by factors such as size, sex, behaviour, density, habitat characteristics, season, tide and time of day. For example, damselfishes, butterfly fishes and angelfishes travel <0.1–0.5 km, while some sharks and tuna migrate over thousands of kilometres. Larval dispersal distances tend to be <5–15 km, and self-recruitment to new habitat is common.

The review indicated that effective marine reserves are more than twice the size of the home range of focal/target species (in all directions). The presence of effective marine management outside the reserve may allow smaller reserves. Reserve size recommendations apply to the specific habitats of focal species, not the overall size. For example, coral reef species require coral reef habitats rather than open ocean or seagrass beds.

Marine reserve whose boundaries are extensively fished benefit from compact shapes (e.g., squares or circles rather than elongated rectangles). Including whole ecological units (e.g., an offshore reef) can reduce exports where that is desired.

Habitats

Minimum sustainable population sizes have not been determined for most marine populations. Instead, fisheries ecologists use the fraction of unfished stock levels as a proxy. Meta-analyses suggest that maintaining populations above ~37% of those levels generally ensures stable populations, although variations in fishing pressure allow fractions as small as 10% or as large as 40% (to protect species such as sharks and some grouper that have lower reproductive output or slower maturation). Higher fractions of habitat protection may protect areas vulnerable to disturbances such as typhoons or climate change. 20–30% protection can achieve fisheries objectives in areas with controlled fishing pressure and is the minimum level of habitat protection recommended by IUCN-WCPA.

Special Areas

Many fish species congregate to facilitate spawning. Such congregations are spatially and temporally predictable and increase the species' vulnerability to overfishing. Species such as groupers and rabbitfishes travel long distances to congregate for days or weeks. Such gatherings are their only opportunities to reproduce and are crucial to population maintenance. Species such as snappers and parrotfishes congregate in feeding or resting areas. Juveniles may congregate in nursery areas without adults. Such special areas may require only seasonal protections if at other times no vital activities are taking place. Such reserves must be spaced to allow focal species to journey among them. If the location of such special areas is unknown, or is too large to include in a reserve, management approaches such as seasonal capture and sales restrictions may provide some protection. Sea turtle nesting areas, dugong feeding areas, cetacean migratory corridors and calving grounds are examples of other special areas that can be protected seasonally. Other types of special areas include isolated habitats that have unique assemblages and populations, habitats that are important for endemic species and highly diverse areas.

Isolated Populations

Isolated populations (e.g. those on remote atolls) have high conservation value where they harbor endemic species and unique assemblages. A location or population 20–30 km from its nearest neighbor generally qualifies as isolated in the absence of a persistent linking current. Their isolation (low connectivity) requires such areas to be largely self-replenishing. This leaves them less resilient to disturbance. Sustaining their marine species requires a higher fraction of living areas to be protected.

Recovery

Coral reef fish species recovery rates (from e.g. overfishing) depend on their life history and factors such as ecological characteristics, fishing intensity and population size. In the Coral Triangle, species at lower trophic levels that have smaller maximum sizes, faster growth and maturation rates and shorter life spans tend to recover more quickly than species having the opposite characteristics. For example, in the Philippines, populations of planktivores (e.g., fusiliers) and some herbivores (e.g., parrotfishes) recovered in <5–10 years in marine reserves, while predators (e.g., groupers) took 20–40 years. Increased fishing pressure adversely affects recovery rates (e.g., Great Barrier Reef and Papua New Guinea).

Long-term protection allows species with slower recovery rates to achieve and maintain ecosystem health and associated fishery benefits. Permanent protection protects these species over the long-term. Short-term protections do not allow slow-recovering species to reach or maintain stable populations.

In some Coral Triangle countries (e.g., Papua New Guinea and Solomon Islands), short term protections are the most common form of traditional marine resource management. These protections can help address problems at lower trophic levels (e.g., herbivores) or allow spawning to succeed. Other reasons for adopting short-term protections include allowing communities to stockpile resources for feasts or close areas for cultural reasons. Short-term/periodic reserves also may function as partial insurance by enhancing overall ecosystem resilience against catastrophes. Reopened reserves can be protected by management controls that limit the harvest to less than the increase achieved during closure, although at greatly reduced recovery rates.

Resilience

Some habitats and species are better prepared environmental changes or extremes. These include coral communities that handle high sea surface temperature (SST); areas with variable SSTs and carbonate chemistry and areas adjacent to undeveloped low-lying inland areas that coastal habitats can expand into as sea levels rise. Such areas constitute climate change refugia and can potentially better protect biodiversity than more fragile areas. They may also provide fishery benefits, since habitat loss from climate change is a major fishery threat.

Local Practices

Local practices such as overfishing, blast fishing, trawling, coastal development and pollution threaten many marine habitats. These threats decrease ecosystem health and productivity and adversely affect focal and other species. Such practices can also decrease resilience. Some practices that originate beyond reserve boundaries (e.g., runoff) can be mitigated by considering their impacts within broader management frameworks. Areas that are not threatened by such practices and that are adjacent to other unthreatening areas may be better choices for reserves.

Networks

Networks of marine reserves can support both fisheries management and biodiversity conservation. The size, spacing and location of reserves within a network must respect larval dispersal and movement patterns of species that are targeted for protection.

- Design: Existing ecological guidelines for designing networks independently focus on achieving either fisheries, biodiversity or climate change objectives or combinations of fisheries and biodiversity or biodiversity and climate change. These three goals have different implications for network design. The most important are reserve size and protection duration (permanent, long term, short term, or periodic closures).
- Diversity: Maintaining diversity involves protecting all species. Generally this involves protecting adequate examples of each major habitat (e.g., each type of coral reef, mangrove and seagrass community). Resiliency to threats improves when multiple examples of each habitat are protected.
- To address biodiversity or climate change, reserves 4–20 km across are recommended, because they protect larger populations of more species.
- Climate change: Protecting areas that have already proven resilient to ecological changes and are relatively well-protected by other protocols are likely to better survive climate change as well.

- Fisheries: Reserves 0.5–1 km across export more adults and larvae to fished areas, potentially increasing recruitment and stock replenishment there. Such small reserves are common in the Coral Triangle, where they benefited some fisheries.
- Connectivity: Connectivity is the linking of local populations through the (voluntary) dispersal of individuals. Connected reserves are close enough to each other that larvae, juveniles or adults can cross from one to another as their behavior patterns dictate. Connectivity is a key factor in network design, since it allows a disturbed reserve to recover by recruiting individuals from other, potentially overpopulated, reserves. Effective networks spaced reserves at distances of <15 km from each other, with smaller reserves spaced more closely.

  Most coastal fish species have a bipartite life cycle where larvae are pelagic before settling out of the plankton to live on a reef. While these fish travel varying distances during their life history, their larvae have the potential to move tens to hundreds of km, more than the more sedentary adults and juveniles, which have home ranges of <1 m to a few km. Adults and juveniles of some species travel tens to hundreds of kilometers as they mature to reach appropriate habitats (e.g., such as coral reef, mangrove and seagrass habitats) or to migrate to spawning areas. When adults and juveniles leave a marine reserve, they become vulnerable to fishing. However, larvae can generally leave a reserve without elevated risk because of their small size and limited fishery exposure. Effective networks account for the movement patterns of target species at each life cycle stage. Given a strong, consistent current, siting marine reserves upstream increases downstream populations.

Open Ocean Reserves

As of April 2008 no high seas marine reserves had been established. Greenpeace is campaigning for the "doughnut holes" of the western pacific to be declared as marine reserves and for 40 percent of the world's oceans to be so protected.

MARINE PARK

A marine park is a type of marine protected area (MPA). An MPA is a section of the ocean where a government has placed limits on human activity. Marine parks are multiple-use MPAs, meaning they have different zones allowing different types of activities. Marine parks usually allow recreational activities, such as boating, snorkeling, and sport fishing. Most marine parks also include zones for commercial fishing, sometimes called open zones. They may also include no-take zones, which prohibit extractive activities, such as fishing, mining, and drilling.

Marine parks are very similar to local parks on land. They are used by the community and often have facilities to encourage their use. They also face many of the same problems as parks on land: overuse and pollution.

East End Marine Park, U.S. Virgin Islands

East End Marine Park protects the largest barrier reef system in the Caribbean Sea. It encompasses 155 square kilometers (60 square miles) of coral reef, shallow sea, and other marine habitats. East End also protects the eastern end of the island of St. Croix, including about 19 kilometers (12 miles) of coastline. Because East End includes both marine and terrestrial (land) habitats, it is considered a hybrid park.

This area of the U.S. Virgin Islands is biologically diverse. An estimated 400 species of tropical fish live in and around the East End, while 17 species of nesting seabirds rely on the park for food and shelter. Seagrass communities also thrive in East End. Seagrass is an important species, one of the few plants that live directly in the ocean. Sea turtles, manatees, fish, and many species of seaweed depend on the seagrass habitat for survival.

Some of the plant and animal species that call East End home are considered threatened or endangered. Elkhorn and staghorn corals dominate the reefs. Various species of brain coral, lettuce coral, star coral, and starlet coral are also found there. Scientists have recently discovered that the popultions of these animals have been rapidly declining over the last three decades.

The park is also home to endangered green turtles, hawksbill turtles, and leatherback turtles. East End includes a turtle refuge, extending about a mile from St. Croixs primary turtle nesting beaches. The park also includes no-take zones, which are off-limits to fishing and harvesting in order to protect the turtles and other threatened species.

Most of East End Marine Park is made of open zones, where most extractive activity, including commercial fishing, is allowed. The only activity prohibited throughout the park is the removal of coral. Other zones are limited to recreational activities, such as sport fishing, boating, and scuba diving.

Kisite Mpunguti Marine Park, Kenya

Kisite Mpunguti Marine Park, in the Indian Ocean, is located off the south coast of Kenya. The park encompasses a marine ecosystem that includes four small islands surrounded by coral reefs. The three Mpunguti Islands are partially covered by dense equatorial rain forest, while Kisite Island is covered in low grasses.

The islands are not inhabited, and the ecosystems remain fairly pristine. No fishing or other extractive activities are allowed in the park. Transportation to the park is limited. These limitations reduce the threat of overfishing and pollution. More than 250 species inhabit Kisite Mpunguti Marine Park, including angelfish, pufferfish, green sea turtles, hawksbill turtles, dolphins, and humpback whales. Seagrasses and tropical seaweeds also thrive in the park.

Kisite Mpunguti is a major destination for snorkelers and scuba divers. The warm water is clear, and the tropical ecosystems have species that exist nowhere else in the world. One of the most unusual species is the coconut crab, the largest land crab in the world. The legspan of a coconut crab can be up to a meter (3 feet). The dolphins that live in Kisite Mpunguti are also popular with tourists, who visit the park in boats, called dhows, from the mainland.

Tourists and historians also visit the islands for research. The islands are filled with caves, where slaves from Africas mainland were held captive before being shipped overseas.

Marine parks allow visitors to interact with marine species in their native environments, although few animals are as friendly as this Florida manatee trying to hitch a ride on a kayak.

Coral Sea Marine Park

The Coral Sea Marine Park (previously known as the Coral Sea Commonwealth Marine Reserve) is an Australian marine park located in the Coral Sea off the coast of Queensland. The marine park covers an area of 989,836 km² (382,178 sq mi) and is assigned IUCN category IV. It is Australia's largest single marine park and is one of the world's largest protected areas.

The Coral Sea Marine Park lies directly adjacent to and east of the Great Barrier Reef Marine Park.

Conservation Values

Species and Habitat

- Humpback whales during their annual migration along the east coast of Australia.
- Nesting and inter-nesting sites for green turtles.
- Breeding and foraging areas for multiple seabird species including noddies, terns, boobies, frigatebirds, and tropic birds.
- White shark distribution and whale shark aggregation.
- Transient populations of highly migratory pelagic species, including small fish schools, billfish, tuna and sharks.
- The East Australian Current forms in the region and is considered a major pathway for mobile predators such as billfish and tunas. Black marlin undergo seasonal movements into the Queensland Plateau area.

Bioregions and other Features

- Includes three Key Ecological Features: The reefs, cays and herbivorous fish of the Queensland Plateau and the Marion Plateau and the northern extent of the Tasmantid Seamount Chain.
- Heritage values include several historic shipwrecks including three World War II shipwrecks from the Battle of the Coral Sea.
- The reserve represents the full range of seafloor features found in the region, including numerous reefs ranging from Ashmore and Boot Reefs in the north of the region to Cato Island and surrounding reefs in the south. The reserve includes canyons, troughs and plateaux, including Bligh Canyon approximately 200 kilometres off the coast from Lockhart River and the Townsville Trough, which separates the Queensland and Marion Plateaux. The reserve extends into the deeper waters of the Coral Sea Basin in the north, and provides protection for the pinnacles of the northern extent of the Tasmantid seamount chain.
- Six provincial bioregions, 94 depth ranges, and 16 seafloor types are represented in the reserve.

Protection Zones

The Coral Sea Marine Park has been assigned IUCN protected area category IV. However, within the marine park there are multiple protection zones, each zone has an IUCN category and related rules for managing activities to ensure the protection of marine habitats and species.

Table: The following table is a summary of the zoning rules within the Coral Sea Marine Park.

Zone	IUCN	Activities permitted						Total area (km2)
		Vessel transiting	Recreational fishing	Commercial fishing	Commercial aquaculture	Commercial tourism	Mining	
National Park	II	Yes	No	No	No	Excludes fishing, with approval	No	238,400
Habitat Protection	IV	Yes	Yes	Some, with approval	With approval	With approval	No	655,129
Habitat Protection (Reefs)	IV	Yes	Yes	Hand and dropline, with approval	With approval	With approval	No	29,827
Special Purpose (Trawl)	VI	Yes	Yes	Most, with approval	With approval	With approval	No	66,480

Marae Moana

Marae Moana is a multiple-use marine protected area created on 13 July 2017, when the Parliament of the Cook Islands passed a bill creating the largest multiple-use marine protected area in the world at the time of its passage. Marae Moana covers the Cook Islands' entire exclusive economic zone of over 1.9 million square kilometers.

Kevin Iro, an environmentalist, first proposed the idea in 2010. Upon its passage, Iro said, "It's a historic time, particularly because everyone supported it, including our traditional leaders who spearheaded the whole thing." The bill had bipartisan support. The name of the marine park was created in 2014 during a naming competition won by Bouchard Solomono, a student at Tereora College. Some media speculated that this legislation was the most significant since the independence of the Cook Islands.

Iro, who is a co-chair of the Cook Islands Marine Park Steering Committee, which was tasked with Marae Moana's legal designation, said this bill and marine park signifies "the sacredness of how Cook Islanders view our ocean space. It links us to our ancestors – it's more than just the ocean."

The legislation also creates fifteen marine protected areas extending to 50 nautical miles around each island where no large scale commercial fishing or seabed minerals activities are permitted. Fishing and mineral exploration will still exist in the exclusive economic zone, but is to be done sustainably.

The Prime Minister, Henry Puna, said, "Socially, economically and spiritually we must all take care of it – and it is imperative that all those that live and exist both within and beyond its boundaries do recognise and respect its sanctity,".

"We do not only recognise that the ocean brings us revenue in terms of fishery and tourism and potentially sea bed minerals – it also provides us with clean air, clean water, and clean food to nourish and sustain us."

"So this bill aims to sustain our livelihoods by protecting species and ecosystems as well as our cultural heritage that we inherit and pass on to future generations."

Iro was inspired to create this legislation from his experience with the islands and ocean and seeing environmental decay such as pollution and overfishing. "When I moved back here (from New Zealand) about 16 years ago I saw what was happening to the lagoons and reefs and really wanted to protect them for my kids," he said. "I want them to have the same experience I did as a boy growing up."

Biodiversity

The park includes atolls, a volcanic island, upraised limestone islands, and a sand cay. It is home to 136 species of coral, over 600 species of fish, and 21 species of cetacea (whales and dolphins). It is also home to three threatened species of turtles, as well as some tuna species. Marae Moana is also home to five species of reef sharks and eleven pelagic shark species. Marae Moana is also home to many birds and plants as well.

Seabed Mining Controversy

In late 2018, the Cook Islands government began consultations to revise the seabed minerals act, which would facilitate exploration for deep sea mining. An estimated value of US$146 billion worth of manganese nodules are to be found six kilometres down on the ocean floor, particularly around the northern islands. The mining methods used to retrieve the minerals will likely be violently damaging to the local ecosystems.

In August 2019, a ten-year moratorium was proposed on seabed mining in the protected area at the Pacific Islands Forum creating both national and international debate. Due to significant support within the government for seabed mining the bill was not passed. Although any mining activity must be kept to certain areas of the park where impact on the marine ecosystems will be limited, the failure of the bill has drawn criticism due to the unclear determination of what areas of the park can be mined. Technical Director of the Te Ipukarea Society (TIS) Kelvin Passfield displayed concern about the lack of information available regarding potential mining sites, saying "we're not saying no to seabed mining, but it depends on more information than we've got now".

The Cook Islands government has also faced a history of accusations of exploiting its national waters, including selling fishing licenses to foreign companies. Henry Puna, the Prime Minister of the Cook Islands, stated that a motivation to introduce deep-sea mining was a factor in the government's support of the Marae Moana: "We have minerals in our ocean, it's just a matter of coming up with the technology to exploit it but not damage the environment. That's why we're keen to [set up] this marine park first so that it sets the standards for any exploitation: That's why we're doing it".

Jacqueline Evans, a prominent marine biologist and Goldman Environmental Prize Recipient who was openly critical of the government's plans for seabed mining was replaced as director of the park by instruction of the Office of the Prime Minister. Evans was responsible for developing the Marae Moana policy and the Marae Moana Act, and since the Act's passage in 2017 has, amongst other things, worked on raising funding for the park, expanding policy and regulations and developing a plan that designates permitted areas for industrial activities. Her removal as director has attracted local media criticism.

Great Barrier Reef Marine Park

The Great Barrier Reef Marine Park protects a large part of Australia's Great Barrier Reef from damaging activities. It is a vast multiple-use Marine Park which supports a wide range of uses, including commercial marine tourism, fishing, ports and shipping, recreation, scientific research and Indigenous traditional use. Fishing and the removal of artefacts or wildlife (fish, coral, seashells, etc.) is strictly regulated, and commercial shipping traffic must stick to certain specific defined shipping routes that avoid the most sensitive areas of the park. The Great Barrier Reef is the largest and best known coral reef ecosystem in the world. Its reefs, almost 3000 in total, represent about 10 per cent of all the coral reef areas in the world. It supports an amazing variety of biodiversity, providing a home to thousands of coral and other invertebrate species, bony fish, sharks, rays, marine mammals, marine turtles, sea snakes, as well as algae and other marine plants.

Managing Authority

The Great Barrier Reef Marine Park Authority (GBRMPA) is responsible for the care and protection of the Great Barrier Reef Marine Park. It uses a range of tools to manage the marine park including Acts and Regulations, zoning plan, plans of management, Traditional Owner agreements, partnerships, stewardship and best practice, education, research and monitoring and reporting. It issues permits for various forms of use of the marine park, and monitors usage in the park to ensure compliance with rules and regulations associated with the park. GBRMPA is funded by Commonwealth Government appropriations and an environmental management charge levied on the permit-holders' passengers. Currently this is A$6.00 per day per passenger (to a maximum of $16.50 per trip).

Extent

The marine park lies east of the mainland coast of Queensland, starting in the north at Cape York. Its northern boundary is the circle of latitude 10°41'S (running east up to the eastern edge of the Great Barrier Reef at 145°19'33"E), thereby encompassing those few uninhabited Torres Strait Islands that are east of Cape York, south of 10°41'S and north of 11°00'S. The largest of those island are Albany Island (5.9 km² or 2.3 sq mi), Turtle Head Island 12.8 km² or 4.9 sq mi and Trochus Island 2.2 km² or 0.85 sq mi. Further islands are Mai Island 0.25 km² or 0.097 sq mi, Bush Island 0.2 km² or 0.077 sq mi, Tree Islet 0.01 km² or 0.0039 sq mi, Brewis Island 0.05 km² or 0.019 sq mi, and a few unnamed islets.

Protected Area Designation

As of 2016, zones within the marine park has been categorized into the following IUCN protected area categories:

- IUCN Category Ia with an area of 859 square kilometres (332 sq mi) being about 0.2% of the marine park's area.
- IUCN Category II with an area of 114,309 square kilometres (44,135 sq mi) being about 33%.
- IUCN Category IV with an area of 15,027 square kilometres (5,802 sq mi) being about 4%.
- IUCN Category VI with an area of 213,769 square kilometres (82,537 sq mi) being about 62%.

Threats

Crown-of-thorns Starfish

The Crown-of- thorns Starfish (Acanthaster planci), while native to the reef, poses one of its greatest threats when outbreaks occur. Several efforts have been made to identify and control these outbreaks, including the COTSBot developed by Queensland University of Technology.

These outbreaks and subsequent threats have led to an increase in awareness and appearances in mainstream media, as industry and conservation groups try to control the COTS and protect the reef. An infographic explaining the process of identifying and targeting this threat can be seen here.

Outlook for the Reef

The Great Barrier Reef Outlook Report 2014, published every five years, examines the Great Barrier Reef's health, pressures, and likely future. The report is required under Great Barrier Reef Marine Park Act 1975 and aims to provide a regular and reliable means of assessing reef health and management in an accountable and transparent way.

The report finds the greatest risks to the Reef are still climate change, land-based run-off, coastal development, some fishing impacts, illegal fishing and poaching.

Associated Protected Areas

The Great Barrier Reef Coast Marine Park is a marine protected area located in Queensland waters adjoining the Great Barrier Reef. It was created on 5 November 2004 by the amalgamation of the following former marine parks - the Mackay/Capricorn Marine Park, the Townville/Whitsunday Marine Park, the Trinity Inlet/Marlin Coast Marine Park and the Cairns Marine Park, along with the addition of previously unprotected area of coastal water that adjoined the former marine parks. The marine park is intended to complement the adjoining GBRMP by the use of "similar zone objectives, and entry and use provisions." It is managed by the Queensland Parks and Wildlife Service.

SHARK SANCTUARY

A shark sanctuary is an area that forbids commercial fishing operations from targeting and retaining caught sharks, including their fins. The first shark sanctuary was created by Palau in 2009. It was followed by Maldives, Honduras, The Bahamas and Tokelau.

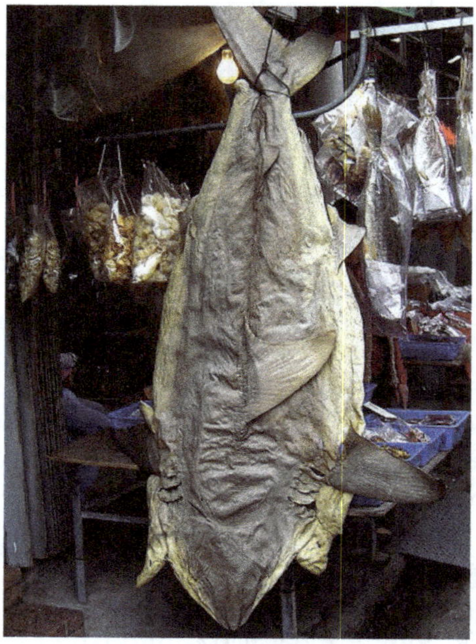

Dried shark skin and fins for sale in market.

Every year, fishermen pull "up to 73 million" or "some 100 million" sharks from the world's oceans. The United Nations Food and Agriculture Organization (FAO) estimates that more than half of shark species are overexploited or depleted. Globally, 21%

of shark species whose extinction risk has been assessed fall into the "threatened" categories, and 18% are "near threatened", while 35% lack sufficient data to decide, leaving 26% in the unthreatened category. The search for shark fins drives the illegal hunting trade. Some jurisdictions permit fishing for fins and food. Sharks are also caught as bycatch when fishing for marlin, tuna and other varieties.

Sharks generally reach sexual maturity only after several years of life and produce very few offspring in comparison to other harvested fish. Harvesting sharks before they reproduce has severe impacts on future populations.

National and International Status

Many nations restrict shark catches and shark finning:

- Palau created the world's first so-named "shark sanctuary" on September 25, 2009. Palau forbids all commercial shark fishing within its Exclusive Economic Zone (EEZ) waters. The sanctuary protects about 600,000 square kilometres (230,000 sq mi) of ocean, an area similar to the country of France. President Johnson Toribiong made the announcement at a meeting of the United Nations. President Toribiong also requested a worldwide ban on shark fishing. Palau is home to 135 endangered or vulnerable shark and ray species.
- The Maldives created a sanctuary in March, 2010. Tokelau declared its entire EEZ a shark sanctuary in 2011.
- On February 25, 2011, Guam, a US island territory, voted to ban commerce in fins. Guam's Senate passed a bill banning the sale, possession and distribution of the fins.
- In August, 2011, Micronesia, the Marshall Islands, and Guam announced plans to join Palau in a region-wide sanctuary that covers 2,000,000 square miles (5,200,000 km^2) of ocean.
- Kiribati, another Pacific island state, and the US operate Earth's largest marine reserve.
- Honduras prohibited the taking of sharks in its national waters as of February 2010.
- The U.S. bans shark finning on all U.S.-flagged vessels, forbids the taking 19 species of sharks including white, whale, and basking sharks and shares lists of illegal vessels with established fishing companies, helping them report illegal activities. The U.S. also assesses the health of many of its shark populations and includes sharks in its various Fishery Management Plans. Hawaii has banned the sale and possession of shark fins. The states of California, Oregon and Washington are considering similar bans.

- In 1991 South Africa became the first country in the world to declare great white sharks a legally protected species.
- In February 2009, the European Commission proposed first-ever shark conservation rules for European waters, although these are not outright bans. EU countries account for one-third of global shark meat exports.
- Shark steaks are increasingly served in restaurants. Shark parts are also used in lotions and leather sports shoes.
- Taiwan banned shark finning in 2012.

Drivers of the Shark Trade

Shark fin soup.

Sharks are a common seafood in many places around the world, including China (shark-fin soup), Japan, Australia (fish and chips under the name flake), in India (under the name sora in Tamil language and Telugu language), and Icelanders eat Greenland sharks as hákarl.

The practice of live finning, where a fisherman removes the fin with a hot metal blade and releases the dying animal occurs with some frequency, but is not a prevalent practice, with most fins taken with the entire animal. Around 80% of shark fins are derived as accidental by-catch, which are quickly sold to the market due to their high price. Shark finning has become a major trade within black markets all over the world with shark fins going at about $300/lb in 2009.

European consumers consume dogfishes, smoothhounds, catsharks, makos, porbeagle and also skates and rays. However, the U.S. FDA recommends that children and women who are or may be pregnant should refrain from eating shark.

In the East Asian region, use of shark cartilage in preparing soups is considered a health tonic. Hong Kong imports it from North and South American countries, particularly for use in either a cooked format or to prepare boiled soup, as a health fad, by mixing it with herbals supplements.

Another large demand for shark cartilage is for manufacture of "Shark Cartilage Powder" or pills as a cure for cancer. The anti-cancer claims of such powders marketed in many parts of the world has been discounted by the US Food and Drug Administration, Federal Trade Commissions and medical studies. In spite of such injunctions, the trade in this powder continues and the shark cartilage powder is still widely marketed as a cancer cure, stated to be selling at US$145 per gram. It is also stated that in Costa Rica, one single firm alone processed 235,000 sharks every month to manufacture cartilage pills.

Seafood Watch recommends that everyone avoid eating shark.

The majority of shark fisheries around the globe have little monitoring or management. With the rise in demand for shark products there is a greater pressure on fisheries. Sharks experience a long interval between birth and sexual maturity, such that many sharks never reach maturity. In some species, populations have declined by over 90% over the past 20–30 years with decline of 70% not unusual.

The practice of shark finning, attracts much controversy and regulations are being enacted to prevent it from occurring. The acclaimed 2007 documentary, Sharkwater exposed how sharks are being hunted to extinction, in part due to the massive Asian demand for shark fin soup.

WHALE SANCTUARIES

The world's first international whale sanctuary was established by the IWC in 1937. This covered one-quarter of the Southern Ocean, millions of square kilometres between New Zealand and the western seaboard of South America. It was disestablished in 1955, in response to demand from an industry facing falling catches elsewhere in the Antarctic. Within a year the former sanctuary was supplying 25 per cent of total Antarctic catch. It remained open till the 1986 moratorium. The serious depletion of great whales in the breeding areas of the South Pacific can be traced to the ending of the Sanctuary, combined with the illegal whaling conducted by the Soviet Union in this area.

In 1979 the IWC agreed to establish the Indian Ocean Whale Sanctuary, protecting whales in breeding and calving grounds in the region. In 1994 the IWC established the Southern Ocean Whale Sanctuary, covering all waters surrounding Antarctica as far northwards as 60 degrees south, and protecting the feeding grounds of three-quarters of the world's whales.

Proposals have been presented at recent IWC meetings for two further sanctuaries: the South Pacific Whale Sanctuary, proposed by Australia and New Zealand, and the South Atlantic Whale Sanctuary, proposed by Brazil and Argentina.

Both would extend the Southern Ocean Whale Sanctuary's coverage from Antarctica to the tropics. Together they would protect Southern Hemisphere whales during their entire life cycles, between feeding in the Antarctic and calving in the tropics.

Neither proposal has yet gained the 75 per cent majority vote required by the IWC. New Zealand will continue to push for the South Pacific Whale Sanctuary in the IWC, in the belief that its establishment would help the recovery of some of the world's most depleted whale populations.

Over the last three years, over 12 million square kilometres of the South Pacific Ocean within their Exclusive Economic Zones have been declared as national whale sanctuaries by a number of Pacific Island states and territories.

Australian Whale Sanctuary

The Australian Whale Sanctuary was established in 1999 to protect dolphins and whales from hunting in waters within the Australian government's jurisdiction.

The sanctuary includes the whole of the Australian Exclusive Economic Zone (EEZ), which is the area 200 nautical miles (370 km) surrounding the continent of Australia and its external dependencies such as Christmas Island (in the Indian Ocean), Cocos (Keeling) Island, Norfolk Island, Macquarie Island and Heard Island and McDonald Islands, and including the EEZ adjoining the coastline of the Australian Antarctic Territory which is only recognised by United Kingdom, New Zealand, France and Norway.

The sanctuary is the scene of an ongoing controversy between Australia and Japan over whaling. In 2008 the Federal Court of Australia ruled it was illegal under Australian law for the Japanese whaling fleet to kill whales in the Sanctuary. In 2015 Kyodo Senpaku Kaisha Ltd was found guilty of wilful contempt of court and fined A$1,000,000.

Great Australian Bight Marine Park Whale Sanctuary

Great Australian Bight Marine Park Whale Sanctuary was a marine protected area in the Australian state of South Australia located immediately off the coastline of the Great Australian Bight in waters about 790 km (491 mi) west-northwest of the state capital of Adelaide.

It extended from 130 degrees 45.5 minutes east at its western boundary to 131 degrees 30 minutes east at its eastern boundary, and from the coastline for a distance of 3 nautical miles (5.6 km; 3.5 mi). It was bounded by the Great Australian Bight Marine National Park to its immediate west and east and its extent included the bay known as the Head of Bight. It was also part of the group of marine protected areas which are located together in waters within Australian and South Australian jurisdictions within the Great Australian Bight and which is collectively known as the Great Australian Bight Marine Park. Since late 2012, both the whale sanctuary and the national park have also been within the boundaries of the Far West Coast Marine Park.

The whale sanctuary was originally proclaimed as a marine park under the Fisheries Act 1982 (SA) by the South Australian Government on 22 June 1995 to protect waters used by Southern right whales for breeding and calving. It was re-classified as an aquatic reserve after the repeal of the Fisheries Act 1982 by the Fisheries Management Act 2007 (SA) in 2007. It was abolished on 20 October 2016.

The whale sanctuary was classified as an IUCN Category Ia protected area.

Southern Ocean Whale Sanctuary

The Southern Ocean Whale Sanctuary is an area of 50 million square kilometres surrounding the continent of Antarctica where the International Whaling Commission (IWC) has banned all types of commercial whaling. To date, the IWC has designated two such sanctuaries, the other being the Indian Ocean Whale Sanctuary.

The northern boundary of the sanctuary follows the 40 °S parallel of latitude except in the Indian Ocean sector where it joins the southern boundary of the Indian Ocean Whale Sanctuary at 55°S, and around South America and into the South Pacific where the boundary is at 60 °S.

The Southern Ocean Whale Sanctuary was established by the IWC in 1994 with 23 countries supporting the agreement and Japan opposing it.

The status of the Southern Ocean Sanctuary is reviewed and open to change by the IWC every 10 years. During the 2004 meeting a proposal was made by Japan to remove the sanctuary, but it failed to reach the 75% majority required (it received 25 votes in favour and 30 votes against with two abstentions).

Controversy and Enforcement

While commercial whaling is prohibited in the Southern Ocean Whale Sanctuary, Japan's Institute of Cetacean Research (ICR) has continued to hunt whales inside the Sanctuary in accordance with a provision in the IWC charter permitting whaling for the purposes of scientific research. Japan clarified its position on whaling in a number of forums, noting that while some whale species are threatened, many such as minke whale are not. Some conservation groups such as the Sea Shepherd Conservation Society claim Japan's whaling operations are conducted in violation of the IWC charter, arguing that Japan's self-imposed annual quota of 1000 whale kills for research is not really scientific activity, but rather a veil to cover commercial fishing operations as whale meat ends up being sold in Japanese markets.

In an open letter to the Japanese government, published in 2002 in the New York Times and sponsored by the World Wildlife Fund (WWF), 21 scientists declared that they "believe Japan's whale 'research' program fails to meet minimum standards for credible science". They were "concerned that Japan's whaling program is not designed to answer scientific questions relevant to the management of whales; that Japan refuses to make the information it collects available for independent review; and that its research program lacks testable hypotheses or other performance indicators consistent with accepted scientific standards". They accused Japan of "using the pretense of scientific research to evade its commitments to the world community".

The Australian delegation to the IWC has argued to repeal the provision that allows nations to harvest whales for scientific research, to no effect.

Japan, meanwhile, lodged a formal objection to the sanctuary with regard to minke whales, meaning that the terms of the sanctuary do not apply to its harvest of that species within the boundaries of the sanctuary. Claiming the Japanese whaling fleet's actions to be illegal, the Sea Shepherd Conservation Society acts against Japanese whaling ships engaged in yearly hunts, attempting to interrupt or cut short the whaling activities. The ICR considers the activists' methods as constituting piracy, terrorism, and illegal harassment of the ICR fleet. Critics of the ICR point out that recent scientific advances allow for the use of non-lethal techniques in cetacean research, such as biopsies or determination of cetacean dietary intake through analysis of DNA samples from whale feces.

During the 2010/2011 whaling season, the ICR was forced to halt its whaling operations due to the actions of the Sea Shepherd Conservation Society, having harvested 172 whales (a fifth of their season quota). The ICR dispatched its whaling fleet as usual to the Northwest Pacific survey (JARPN II), with a quota for 260 whales, including 100 Minkes.

Dispute over Legality

Japan has argued that the establishment of the Southern Ocean Whale Sanctuary was in contravention of the International Convention for the Regulation of Whaling (ICRW) on which the IWC is based and is therefore illegal.

This view received strong support from Professor W. T. Burke of the University of Washington in his paper circulated as IWC Document Number IWC/48/33. He refers to Article V of the ICRW, which states that the creation of any sanctuary must "be based on scientific findings" and "take into consideration the interests of the consumers of whale products and the whaling industry".

Prof Burke's interpretation is disputed by Prof Chris Wold of the Northwestern School of Law of Lewis and Clark College, who argued that the IWC must maintain both the conservation and whaling industry objectives of the ICRW. This view is supported by Alexander Gillespie, Professor of Law at the University of Waikato, who reasoned in his 2006 book, Whaling Diplomacy, that phrases such as "best interests of the whaling industry" must now consider commercial non-consumptive use (such as whale watching).

As there is no settlement procedure in the IWC for this type of dispute, Japan has asked the IWC to submit its case to a relevant legal body for analysis. This proposal was rejected by a majority of IWC members.

International Court of Justice (ICJ) Proceedings

On 31 May 2010, the Australian Government lodged formal proceedings against Japan in the International Court of Justice (ICJ) in The Hague, Netherlands. In a joint ministerial statement, the government stated that it "has not taken this decision lightly."

On 5 January 2011, leaked US diplomatic cables revealed that the Australian legal challenge to Japanese whaling was heavily influenced by domestic political pressures and Australian government advisers were left deeply pessimistic about the prospects of success in the International Court. In its reports to Washington, the US Embassy reported that Australian government contacts were stating that the legal proceedings "would be unlikely to stop the whale hunt entirely, but could well force modifications that would make it more difficult for the Japanese". The Australian public overwhelmingly supports governmental efforts to stop Japanese whaling, even at risk of losing valuable trade relations. The New Zealand government lodged a "Declaration of Intervention" with the ICJ on 6 February 2013, in which it deemed Japan as ineligible for a Special Permit that would allow whaling on the basis of scientific research.

As a result of the Australian government's 2010 application, a court case was heard at the ICJ. The hearing ran from 26 June 2013 until 6 July 2013. In its deposition to the ICJ, the Australian government has claimed that Japan "has breached and is continuing to breach" its obligations under the international convention, and further asserted that Japan has refused to accept International Whaling Commission (IWC) recommendations. Solicitor-General of Australia Justin Gleeson appeared for Australia before the ICJ.

The ICJ's 16-judge bench received and then decided upon the case. Rothwell stated that the case "is a complex one which raises novel legal issues not previously considered by any international court".

Tony Press, CEO of the Antarctic Climate and Ecosystems Cooperative Research Centre at the University of Tasmania, Australia, provided weekly reports of the ICJ hearings for The Conversation Australia media publication. Following the second week, Press relayed that Japan articulated in certain terms its intentions in the Southern Ocean: "the proper conservation of whale stocks and thus make possible the orderly development of the whaling industry", further explaining that a whale must be killed to obtain certain types of information, such as the collection of ear plugs to estimate the age of a whale. During the second week, Professor Lars Walløe, of the University of Oslo and a former Norwegian Commissioner to the Whaling Commission was presented by Japan as its sole expert witness.

During the week beginning 8 July 2013, New Zealand delivered its intervention, in which it provided a history of the origin of the 1946 Whaling Convention and Australian Attorney-General Mark Dreyfus affirmed that his nation had not "colluded" with New Zealand to launch the ICJ proceedings. Dreyfus concluded Australia's contribution by emphasising the friendship shared by Australia and Japan, and stated: "Australia respectfully requests the Court to bring Japan's whaling program to an end."

The ICJ hearing concluded on 16 July 2013. In Japan's final presentation Professor Payam Akhavan of Harvard University stated: "It would not be an exaggeration to say that Australia's case now hangs by a thread".

Outcome

On 31 March 2014 the ICJ ruled that Japan's JARPA II whaling program "can broadly be characterized as "scientific research", but failed to comply with three provisions of the Schedule to the International Convention for the Regulation of Whaling. The Court ordered that "Japan revoke any extant authorization, permit or licence to kill, take or treat whales in relation to JARPA II, and refrain from granting any further permits under Article VIII, paragraph 1, of the Convention, in pursuance of that programme". The court's judges agreed with Australia that the number of Japanese scientific research publications were not proportionate to the number of animals killed. Two former Australian Environment ministers applauded the decision and stated their hopes that Japan will respect it. The decision of the ICJ is final and the right of appeal does not apply in this context; Japan said it would abide by the decision.

Indian Ocean Whale Sanctuary

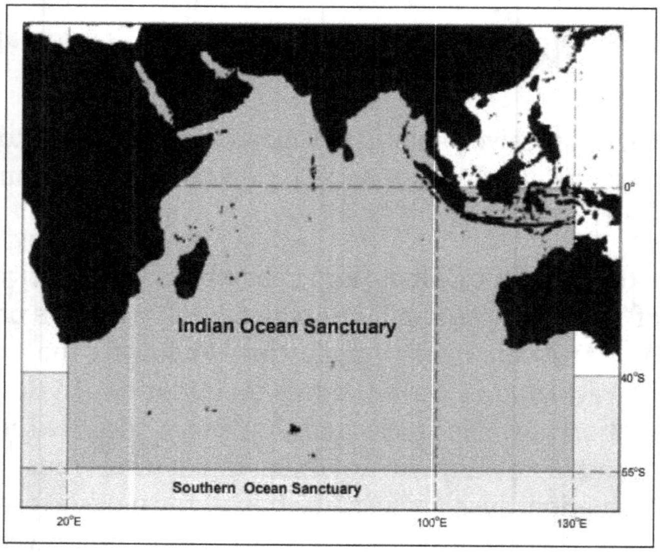

From the unusual pygmy blue whales off the West coast of Sri Lanka, to the humpback whales off the South Coast of Oman, the Indian Ocean is an important area for cetaceans, large and small. Unlike the Atlantic and Pacific oceans, the Indian Ocean has a Northern coastline, which prevents the migration of whales to Arctic waters and may have led to the evolution of unique populations of whales.

Indian Ocean Whaling

Though many populations of whales were devastated by commercial whaling, most survive and the Sanctuary provides a long-term opportunity for their eventual recovery. The most remarkable story is that of Oman's humpback whales, which, unknown to the world, were almost wiped out by illegal Soviet whalers in the 1960s. Following the collapse of the Soviet Union, efforts were made to document the illegal whaling of the Soviets, involving several of the scientists who were present on the whaling ships at the time. It is now clear that Oman's humpback whales survived – but only just.

In the 19th Century, whalers took large numbers of sperm whales – sadly, almost exterminating them from some areas, such as around the Seychelles, the country responsible for the original sanctuary proposal.

For many years the IWC's ban on factory ship whaling for baleen whales north of 40°S gave partial protection for whales in the Indian Ocean, but as whales in the Antarctic became scarcer, Japanese scouting vessels began to prospect in the Indian Ocean. Between 1977-79, Japan caught hundreds of Bryde's whales in the Indian Ocean for supposedly scientific purposes, and repeatedly tried to have the restrictions on factory ship whaling lifted, until the Sanctuary came into effect in 1979. Plans to establish land-based whaling stations in Indian Ocean countries were shelved when the Sanctuary was adopted.

From the 1960's until 1978, thousands of sei whales were taken by factory ships in the Indian Ocean Sanctuary area near the southern boundary at 55 °S. The fleets of the USSR in the 1970's also took thousands of sperm whales, including females and family groups, in the Indian Ocean north of 40 °S until the Sanctuary came into effect in 1979.

Supported by the Indian Ocean Range States

In 1979, the Seychelles, which proposed the Sanctuary, came with messages of support from many Indian Ocean range states, which were not then members of the IWC. India, Kenya, Sri Lanka and Mauritius are mentioned in the Seychelles proposal. During the meeting the Seychelles spoke of others that indicated support - Somalia, Tanzania and Bangladesh.

Most recently, the Nairobi Convention, at its third Conference of the Parties, hosted in Maputo 5-7 December 2001 pledged its support for the Sanctuary. The meeting "Reaffirm the need to maintain the status of the Indian Ocean as a sanctuary for the protection of endangered marine mammals of the convention region". The Nairobi Convention is under the auspices of UNEP and is the Convention for the Protection, Management and

Development of the Marine and Coastal Environment in the Eastern African Region. Comoros, France (Reunion), Kenya, Madagascar, Mauritius, Mozambique, Seychelles, Somalia, South Africa and Tanzania are members.

Special Support

The USA and France seconded the 1979 proposal by Seychelles. France remains firmly behind the concept of sanctuaries and sponsored a proposal for a sanctuary in the Southern Ocean, which was adopted by the IWC in 1994.

The Indian Ocean Sanctuary

Although there was some opposition in 1979, only three countries voted against the Sanctuary and in the two subsequent reviews there was consensus within the IWC. Of the three who voted against (USSR, Japan and Korea), none are range states to the Sanctuary. Japan now argues against sanctuaries as contrary to the IWC Convention. However, Japan accepted, in principle, the 1994 decision to declare the Southern Ocean a sanctuary by only objecting to the inclusion of minke whales in the decision and did not file an objection to the Indian Ocean Sanctuary.

The Sanctuary was adopted for an initial ten-year period with the provision for a general review after five years. In 1989 the Sanctuary was extended for three years and in 1992 the Sanctuary was extended indefinitely, but with a review in 2002. Both the 1989 and the 1992 decisions were by consensus.

Japan did not file a formal objection to the initial decision and therefore is bound by it. Japan also remained part of the consensus for the 1989 renewal and the 1992 decision that the Sanctuary be adopted indefinitely.

The IWC and Sanctuaries

The IWC is the global body responsible for the conservation of whales and the regulation of whaling. The IWC has over 40 members, though some are behind with their annual dues and therefore cannot vote. Article V of the IWC Convention allows the IWC to designate sanctuary areas.

The Southern Ocean, consisting of most of the waters South of 40 °S, was declared a sanctuary by the IWC in 1994. The Indian and Southern Ocean sanctuaries are contiguous. One of the arguments in favour of the Southern Ocean Sanctuary was to protect the Indian Ocean's whales when they migrated south to feed in the waters around the Antarctic.

References

- Marine-protected-area, topics-earth-and-planetary-sciences: sciencedirect.com, Retrieved 18 June, 2020

- Carleton., Ray, G. (2014). Marine conservation : science, policy, and management. McCormick-Ray, Jerry. Hoboken, NJ: Wiley-Blackwell. ISBN 9781118714430. OCLC 841199230
- White, A. T.; Green, A. L. (2014). "Introduction". Coastal Management. 42 (2): 81–86. doi:10.1080/08920753.2014.877758
- Marine-park, encyclopedia: nationalgeographic.org, Retrieved 17 February, 2020
- Great Australian Bight Marine Park (Commonwealth Waters) Management Plan 2005-2012. Director of National Parks, Commonwealth of Australia. 2005. p. 3. ISBN 0 642 550794

4

Wildlife Sanctuaries

Wildlife sanctuaries refer to the natural habitat that protects the wildlife species from hunting, predation and poaching. It includes bird reserves, animal sanctuaries, protected forests, etc. The topics elaborated in this chapter will help in gaining a better perspective of wildlife sanctuaries.

A wildlife sanctuary is an area where animal habitats and their surroundings are protected from any sort of disturbance. The capturing, killing and poaching of animals are strictly prohibited in these regions. They aim at providing a comfortable living to the animals.

Importance of Wildlife Sanctuaries

There are a number of reasons for establishing wildlife sanctuaries. Some of the reasons are listed below:

- The wildlife sanctuaries are established to protect the endangered species.
- It is quite difficult to always relocate the animals from their natural habitat, therefore, protecting them in their natural environment is advantageous.
- The endangered species are specially monitored in the wildlife sanctuaries. If they reproduce and grow in number while under protection, few specimens can be kept for breeding in the conservation parks for their survival.
- Biologist activities and researches are permitted in the wildlife sanctuaries so that they can learn about the animals living there.
- A few sanctuaries take in injured and abandoned animals and rehabilitate them to health before releasing them in the forest.
- Wildlife sanctuaries preserve the endangered species and protect them from humans and predators.

Protection of Endangered Species

Many species of plants and animals are on the verge of extinction. Such creatures are conserved in the wildlife sanctuaries. Various sanctuaries have been established such as the Fleurieu peninsula sanctuary is maintained to protect sheoak habitat for glossy black cockatoos.

Conservation of Biodiversity

The landowners of a wildlife sanctuary are involved in:

- Production and distribution of electricity.
- Horticulture, grazing and farming enterprises.
- Wine production.
- Organic horticulture.

Ecotourism

Many sanctuaries are involved in ecotourism. They offer accommodation, tour guides, nature walks, etc.

Wildlife Conservation

Ankeny Wildlife Refuge in Oregon.

Wildlife conservation is the practice of protecting wild species and their habitats in order to prevent species from going extinct. Major threats to wildlife include habitat destruction/degradation/fragmentation, overexploitation, poaching, pollution and climate change. The IUCN estimates that 27,000 species of the ones assessed are at risk for extinction. Expanding to all existing species, a 2019 UN report on biodiversity put this estimate even higher at a million species. It's also being acknowledged that an increasing number of ecosystems on Earth containing endangered species are disappearing. To address these issues, there have been both national and international governmental efforts to preserve Earth's wildlife. Prominent conservation agreements include the 1973 Convention on International Trade in Endangered Species of Wild Fauna and Flora (CITES) and the 1992 Convention on Biological Diversity (CBD). There are also numerous nongovernmental organizations (NGO's) dedicated to conservation such as the Nature Conservancy, World Wildlife Fund, and Conservation International.

Species Conservation

Leatherback sea turtle (Dermochelys coriacea).

It's estimated that, because of human activities, current species extinction rates are about 1000 times greater than the background extinction rate (the 'normal' extinction rate that occurs without additional influence). According to the IUCN, out of all species assessed, over 27,000 are at risk of extinction and should be under conservation. Of these, 25% are mammals, 14% are birds, and 40% are amphibians. However, because not all species have been assessed, these numbers could be even higher. A 2019 UN report assessing global biodiversity extrapolated IUCN data to all species and estimated that 1 million species worldwide could face extinction. Yet, because resources are limited, sometimes it's not possible to give all species that need conservation due consideration. Deciding which species to conserve is a function of how close to extinction a species is, whether the species is crucial to the ecosystem it resides in, and how much we care about it.

Leatherback Sea Turtle

The leatherback sea turtle (Dermochelys coriacea) is the largest turtle in the world, is the only turtle without a hard shell, and is endangered. It is found throughout the central Pacific and Atlantic Oceans but several of its populations are in decline across the globe. The leatherback sea turtle faces numerous threats including being caught as bycatch, harvest of its eggs, loss of nesting habitats, and marine pollution. In the US where the leatherback is listed under the Endangered Species Act, measures to protect it include reducing bycatch captures through fishing gear modifications, monitoring and protecting its habitat (both nesting beaches and in the ocean), and reducing damage from marine pollution. There is currently an international effort to protect the leatherback sea turtle.

Habitat Conservation

Habitat conservation is the practice of protecting a habitat in order to protect the species within it. This is sometimes preferable to focusing on a single species especially if the species in question has very specific habitat requirements or lives in a habitat with many other endangered species. The latter is often true of species living in biodiversity hotspots, which are areas of the world with an exceptionally high concentration of endemic species (species found nowhere else in the world). Many of these hotspots are in the tropics, mainly tropical forests like the Amazon. Habitat conservation is usually carried out by setting aside protected areas like national parks or nature reserves. Even when an area isn't made into a park or reserve, it can still be monitored and maintained.

Red-cockaded woodpecker (Picoides borealis).

Red-cockaded Woodpecker

The red-cockaded woodpecker (Picoides borealis) is an endangered bird in the southeastern US. It only lives in longleaf pine savannas which are maintained by wildfires in mature pine forests. Today, it is a rare habitat (as fires have become rare and many pine forests have been cut down for agriculture) and is commonly found on land occupied by US military bases, where pine forests are kept for military training purposes and occasional bombings (also for training) set fires that maintain pine savannas. Woodpeckers live in tree cavities they excavate in the trunk. In an effort to increase woodpecker numbers, artificial cavities (essentially birdhouses planted within tree trunks) were installed to give woodpeckers a place to live. An active effort is made by the US military and workers to maintain this rare habitat used by red-cockaded woodpeckers.

Conservation Genetics

Conservation genetics studies genetic phenomena that impact the conservation of a species. Most conservation efforts focus on ensuring population growth but genetic diversity also greatly affect species survival. High genetic diversity increases survival because it means greater capacity to adapt to future environmental changes. Meanwhile, effects associated with low genetic diversity, such as inbreeding depression and loss of diversity from genetic drift, often decrease species survival by reducing the species' capacity to adapt or by increasing the frequency of genetic problems. Though not always the case, certain species are under threat because they have very low genetic diversity. As such, the best conservation action would be to restore their genetic diversity.

Florida panther (Puma concolor coryi).

Florida Panther

The Florida panther is a subspecies of puma (specifically Puma concolor coryi) that resides in the state of Florida and is currently endangered. Historically, the Florida panther's range covered the entire southeastern US. In the early 1990s, only a single population with 20-25 individuals were left. The population had very low genetic diversity, was highly inbred, and suffered from several genetic issues including kinked tails, cardiac defects, and low fertility. In 1995, 8 female Texas pumas were introduced to the Florida population. The goal was to increase genetic diversity by introducing genes from a different, unrelated puma population. By 2007, the Florida panther population had tripled and offspring between Florida and Texas individuals had higher fertility and less genetic problems. In 2015, the US Fish and Wildlife Service estimated there were 230 adult Florida panthers and in 2017, there were signs that the population's range was expanding within Florida.

Conservation Methods

Wildlife Population Monitoring

Monitoring of wildlife populations is an important part of conservation because it allows managers to gather information about the status of threatened species and to measure the effectiveness of management strategies. Monitoring can be local, regional, or range-wide and can include one or many distinct populations. Metrics commonly gathered during monitoring include population numbers, geographic distribution, and genetic diversity among others.

Monitoring methods can be categorized as either "direct" or "indirect". Direct methods rely on directly seeing or hearing the animals, whereas indirect methods rely on "signs" that indicate the animals are present. For terrestrial vertebrates, common direct monitoring methods include direct observation, mark-recapture, transects, and variable plot surveys. Indirect methods include track stations, fecal counts, food removal, open or closed burrow-opening counts, burrow counts, runaway counts, knockdown cards, snow tracks, or responses to audio calls.

For large, terrestrial vertebrates, a popular method is to use camera traps for population estimation with mark-recapture techniques. This method has been used successfully with tigers, black bears, and numerous other species. Mark-recapture methods are also used with genetic data from non-invasive hair or fecal samples. Such information can be analyzed independently or in conjunction with photographic methods to get a more complete picture of population viability.

Government Involvement

In the US, the Endangered Species Act of 1973 was passed to protect US species deemed in danger of extinction. The concern at the time was that the country was losing species that were scientifically, culturally, and educationally important. In the same year, the Convention on International Trade in Endangered Species of Fauna and Flora (CITES) was passed as part of an international agreement to prevent the global trade of endangered wildlife. In 1980, the World Conservation Strategy was developed by the IUCN with help from the UN Environmental Programme, World Wildlife Fund, UN Food and Agricultural Organization, and UNESCO. Its purpose was to promote the conservation of living resources important to humans. In 1992, the Convention on Biological Diversity (CBD) was agreed on at the UN Conference on Environment and Development (often called the Rio Earth Summit) as an international accord to protect the Earth's biological resources and diversity.

According to the National Wildlife Federation, wildlife conservation in the US gets a majority of its funding through appropriations from the federal budget, annual federal and state grants, and financial efforts from programs such as the Conservation Reserve Program, Wetlands Reserve Program and Wildlife Habitat Incentives Program. A substantial amount of funding comes from the sale of hunting/fishing licenses, game tags, stamps, and excise taxes from the purchase of hunting equipment and ammunition.

Non-government Involvement

In the late 1980s, as the public became dissatisfied with government environmental conservation efforts, people began supporting private sector conservation efforts which included several non-governmental organizations (NGOs). Seeing this rise in support for NGOs, the U.S. Congress made amendments to the Foreign Assistance Act in 1979 and 1986 "earmarking U.S. Agency for International Development (USAID) funds for biodiversity". From 1990 till now, environmental conservation NGOs have become increasingly more focused on the political and economic impact of USAID funds dispersed for preserving the environment and its natural resources. After the terrorist attacks on 9/11 and the start of former President Bush's War on Terror, maintaining and improving the quality of the environment and its natural resources became a "priority" to "prevent international tensions" according to the Legislation on Foreign Relations Through and section 117 of the 1961 Foreign Assistance Act.

Non-governmental Organizations

Many NGOs exist to actively promote, or be involved with, wildlife conservation:

- The Nature Conservancy is a US charitable environmental organization that works to preserve the plants, animals, and natural communities that represent the diversity of life on Earth by protecting the lands and waters they need to survive.
- World Wide Fund for Nature (WWF) is an international non-governmental organization working on the issues regarding the conservation, research and restoration of the environment, formerly named the World Wildlife Fund, which remains its official name in Canada and the United States. It is the world's largest independent conservation organization with over 5 million supporters worldwide, working in more than 90 countries, supporting around conservation and environmental projects around the world. It is a charity, with approximately 60% of its funding coming from voluntary donations by private individuals. 45% of the fund's income comes from the Netherlands, the United Kingdom and the United States.
- Conservation International.
- Wild Team.
- Wild life Conservation Society.
- Audubon Society.
- Traffic (conservation programme).
- Born Free Foundation.
- African Wildlife Defence Force.
- Save Cambodia's Wildlife.
- Wild Earth Guardians.

WILDLIFE MANAGEMENT

Wildlife management attempts to balance the needs of wildlife with the needs of people using the best available science. Wildlife management can include game keeping, wildlife conservation and pest control. Wildlife management draws on disciplines such as mathematics, chemistry, biology, ecology, climatology and geography to gain the best results.

Wildlife conservation aims to halt the loss in the Earth's biodiversity by taking into consideration ecological principles such as carrying capacity, disturbance and succession and environmental conditions such as physical geography, pedology and hydrology. Most wildlife biologists are concerned with the preservation and improvement of habitats although rewilding is increasingly being used. Techniques can include reforestation, pest control, nitrification and denitrification, irrigation, coppicing and hedge laying.

Game keeping is the management or control of wildlife for the well-being of game and may include the killing of other animals which share the same niche or predators to maintain a high population of more profitable species, such as pheasants introduced into woodland. In his 1933 book Game Management, Aldo Leopold, one of the pioneers of wildlife management as a science, defined it as "the art of making land produce sustained annual crops of wild game for recreational use".

Pest control is the control of real or perceived pests and can be used for the benefit of wildlife, farmers, game keepers or human safety. In the United States, wildlife management practices are often implemented by a governmental agency to uphold a law, such as the Endangered Species Act.

In the United Kingdom, wildlife management is undertaken by several organizations including government bodies such as the Forestry Commission, Charities such as the RSPB and The Wildlife Trusts and privately hired gamekeepers and contractors. Legislation has also been passed to protect wildlife such as the Wildlife and Countryside Act 1981. The UK government also give farmers subsidies through the Countryside Stewardship Scheme to improve the conservation value of their farms.

Types of Wildlife Management

There are two general types of wildlife management:

- Manipulative management acts on a population, either changing its numbers by direct means or influencing numbers by the indirect means of altering food supply, habitat, density of predators, or prevalence of disease. This is appropriate when a population is to be harvested, or when it slides to an unacceptably low density or increases to an unacceptably high level. Such densities are inevitably the subjective view of the land owner, and may be disputed by animal welfare interests.

- Custodial management is preventive or protective. The aim is to minimize external influences on the population and its habitat. It is appropriate in a national park where one of the stated goals is to protect ecological processes. It is also appropriate for conservation of a threatened species where the threat is of external origin rather than being intrinsic to the system. Feeding of animals by visitors is generally discouraged.

Opposition

The control of wildlife through killing and hunting has been criticized by animal rights and animal welfare activists. Critics object to the real or perceived cruelty involved in some forms of wildlife management. They also argue against the deliberate breeding of certain animals by environmental organisations—who hunters pay money to kill—in pursuit of profit. Additionally, they draw attention to the attitude that it is acceptable to kill animals in the name of ecosystem or biodiversity preservation, yet it is seen as unacceptable to kill humans for the same purpose; asserting that such attitudes are a form of discrimination based on species-membership i.e. speciesism.

Environmentalists have also opposed hunting where they believe it is unnecessary or will negatively affect biodiversity. Critics of game keeping note that habitat manipulation and predator control are often used to maintain artificially inflated populations of valuable game animals (including introduced exotics) without regard to the ecological integrity of the habitat.

Game keepers in the UK claim it to be necessary for wildlife conservation as the amount of countryside they look after exceeds by a factor of nine the amount in nature reserves and national parks.

Management of Hunting Seasons

Wildlife management studies, research and lobbying by interest groups help designate times of the year when certain wildlife species can be legally hunted, allowing for surplus animals to be removed. In the United States, hunting season and bag limits are determined by guidelines set by the United States Fish and Wildlife Service for migratory game such as waterfowl and other migratory gamebirds. The hunting season and bag limits for state regulated game species such as deer are usually determined by State game Commissions, which are made up of representatives from various interest groups, wildlife biologists, and researchers. Open and closed season on deer in the UK is legislated for in the Deer Act 1991 and the Deer Act 1996.

Open Season

Open season is when wildlife is allowed to be hunted by law and is usually not during the breeding season. Hunters may be restricted by sex, age or class of animal, for instance there may be an open season for any male deer with 4 points or better on at least one side.

Limited Entry

Where the number of animals taken is to be tightly controlled, managers may have a type of lottery system called limited. Many apply, few are chosen. These hunts may still have age, sex or class restrictions.

Closed Season

Closed season is when wildlife is protected from hunting and is usually during its breeding season. Closed season is enforced by law, any hunting during closed season is punishable by law and termed as illegal hunting or poaching.

Type of Weapon Used

In wildlife management one of the conservation principles is that the weapon used for hunting should be the one that causes the least damage to the animal and is sufficiently effective so that it hits the target. Given State and Local laws, types of weapon can also vary depending on type, size, sex of game and also the geographical layout of that specific hunting area.

WILDLIFE MANAGEMENT AREAS

Wildlife management areas (WMAs) are part of outdoor recreation system and are established to protect those lands and waters that have a high potential for wildlife production, public hunting, trapping, fishing, and other compatible recreational uses. They are the backbone to DNR's wildlife management efforts and are key to:

- Protecting wildlife habitat for future generations.
- Providing citizens with opportunities for hunting, fishing and wildlife watching.
- Promoting important wildlife-based tourism in the state.

How did WMAs get Started?

WMA system started in 1951, when the State established its "Save the Wetlands" program to buy wetlands and other habitats from willing sellers to address the alarming loss of wildlife habitat in the state. Initiated by a handful of visionary wildlife managers, the WMA program evolved into the present-day system of WMAs.

How many WMAs are there and Where are they Located?

As a result of more than 50 years of support by hunters, trappers, wildlife enthusiasts, and legislators, today there are over 1.3 million acres of high quality habitat in about 1,500 WMAs located throughout the state, making it one of the best and largest WMA systems in the country.

Who uses WMAs?

Ranging from prairies and wetlands to forests and brushlands, WMAs provide opportunities for hunting, fishing, trapping, and wildlife watching activities. Hundreds of thousands of hunters use these public wildlife lands. Pheasants, waterfowl, deer, and ruffed grouse are the major game species hunted, but WMAs also provide wild turkey, sharp-tailed grouse, rabbit, and squirrel hunting. Wildlife resources are very important.

Management of WMAs

WMAs are the backbone of DNR's wildlife management efforts. Much of the wildlife managers' work is directed toward protecting and enhancing wildlife habitat on WMA lands. For instance, prairie and grasslands are planted to provide prime nesting cover critical to waterfowl and pheasant production. Wetlands are restored and enhanced to benefit waterfowl and other wetland wildlife species. Pheasants also find excellent winter cover in cattails and other marsh vegetation. Prescribed burning is done to maintain grasslands, prairies, and brush lands is important to sharp-tailed grouse and prairie chickens. Forest openings and regeneration projects benefit ruffed grouse, wild turkeys, deer, and moose. Wildlife food plots are managed to feed both resident and migratory wildlife. Woody shelter belts are planted to provide winter cover and nesting sites for upland birds and a variety of nongame species as well.

Who Pays to Acquire WMAs?

Legislature and sportsmen have funded WMA land acquisition in a multitude of different ways. The mainstay of funding has been the surcharge on the small game hunting license. Hunting license fees, bonding funds, Reinvest funds, including Critical Habitat License Plate dollars, and Environmental and Natural Resources Trust Fund (ETF) funds have also been used to buy WMAs. Conservation groups also donate land and money to support the acquisition of WMA lands. Another major source of WMA acquisition funding available to DNR and private conservation partners is the Outdoor Heritage Fund. The Outdoor Heritage Fund is one of several created by the Clean Water, Land and Legacy Amendment to the constitution in 2008.

Future Outlook for WMAs

Despite the program's success, the state is still losing valuable wildlife habitat at an alarming rate. Continued management efforts on existing WMA lands and acquisition of new parcels will be critical to maintaining quality wildlife habitat. According to a 2002 Citizen's Advisory Committee Report on the direction the WMA system should take, acquisition efforts should be accelerated with a long-term 50-year goal of acquiring 702,200 acres of new WMA lands.

IMPORTANT BIRD AREA

An Important Bird and Biodiversity Area (IBA) is an area identified using an internationally agreed set of criteria as being globally important for the conservation of bird populations.

IBA at Guandu Nature Park.

IBA was developed and sites are identified by BirdLife International. Currently there are over 12,000 IBAs worldwide. These sites are small enough to be entirely conserved and differ in their character, habitat or ornithological importance from the surrounding habitat. In the United States the Program is administered by the National Audubon Society.

Often IBAs form part of a country's existing protected area network, and so are protected under national legislation. Legal recognition and protection of IBAs that are not within existing protected areas varies within different countries. Some countries have a National IBA Conservation Strategy, whereas in others protection is completely lacking.

Criteria

Frank Lake (South) IBA.

IBAs are determined by an internationally agreed set of criteria. Specific IBA thresholds are set by regional and national governing organizations. To be listed as an IBA, a site must satisfy at least one of the following rating criteria:

- Globally threatened species: The site qualifies if it is known, estimated or thought to hold a population of a species categorized by the IUCN Red List as Critically Endangered, Endangered or Vulnerable. In general, the regular presence of a Critical or Endangered species, irrespective of population size, at a site may be sufficient for a site to qualify as an IBA. For Vulnerable species, the presence of more than threshold numbers at a site is necessary to trigger selection.
- Restricted-range species: The site forms one of a set selected to ensure that all restricted-range species of an Endemic Bird Area (EBA) or a Secondary Area (SA) are present in significant numbers in at least one site and preferably more.
- Biome-restricted species: The site forms one of a set selected to ensure adequate representation of all species restricted to a given biome, both across the biome as a whole and for all of its species in each range state.
- Congregations:
    - This applies to 'waterbird' species as defined by Delaney and Scott and is modelled on criterion 6 of the Ramsar Convention for identifying wetlands of international importance. Depending upon how species are distributed, the 1% thresholds for the biogeographic populations may be taken directly from Delaney and Scott, they may be generated by combining flyway populations within a biogeographic region or, for those for which no quantitative thresholds are given, they are determined regionally or inter-regionally, as appropriate, using the best available information.
    - This includes those seabird species not covered by Delaney and Scott. Quantitative data are taken from a variety of published and unpublished sources.
    - This is modelled on criterion 5 of the Ramsar Convention for identifying wetlands of international importance. The use of this criterion is discouraged where quantitative data are good enough to permit the application of A4i and A4ii.
    - The site is known or thought to exceed thresholds set for migratory species at bottleneck sites.

The assessment by expert individuals is however not entirely reliable and a study in South America found that the coverage needed for at-risk bird conservation as chosen by computational algorithms rarely overlapped with IBAs and suggested that such methods should be used to complement expert driven IBA site choices.

BIRD RESERVE

A bird reserve (also called ornithological reserve) is a wildlife refuge designed to protect bird species. Like other wildlife refuges, the main goal of a reserve is to prevent species from becoming endangered or extinct. Typically, bird species in a reserve are protected from hunting and habitat destruction. Because of the protection they provide from such threats, bird reserves also serve as excellent locations for bird watching. Normally, wildlife refuges are under the care of non-profit organizations and governmental institutions.

Sign marking the borders of the bird reserve.

Protected Area

A bird reserve is an area where steps have been taken to preserve a habitation that is vital to birds. It can be areas such as wetlands, that normally attract many birds, or land that is important to the survival of an endangered species. Globally, there are different kinds of protected areas with varying legal status. Examples of such are international protection programs, programs governed by the European Union, national and regional programs.

Protection Programs

- IBA: IBA (Important Bird Area) is an international protection program initiated by BirdLife International. An IBA area is a habitat that is vital for protecting birds from a global perspective. In 2012, there were over 10,000 IBA areas around the world. The areas are identified by BirdLife International and have to be small enough so that the entire habitat can be preserved. The area must also be of a different character, easily distinguished, than the surrounding habitats.

- The IBA areas are often part of already existing protected areas and are therefore under protection of national law. Legal status and protection for areas that are not part of national protected area, vary depending on the country.

- SPA: The European Union's Birds Directive requires that special protected areas for birds, so called SPA (Special Protection Areas).
- Ramsar: The Ramsar Convention (The Convention on Wetlands of International Importance, especially as Waterfowl Habitat) is an international agreement to protect wetlands, with special focus on aquatic birds.

The nature reserve Wicken Fen.

Around the World

In Sweden it is the County administrative boards of Sweden and the Environmental Protection Agency that selects which areas are to be designated as bird reserves. Among the noted areas are Tåkern, Lake Hornborga, the southern part of Öland and Getterön. There are approximately 50 Ramsar areas in Sweden, and around 530 SPA areas where the latter comprises almost 30,000 km^2 (12,000 sq mi).

ANIMAL SANCTUARY

An animal sanctuary is a facility where animals are brought to live and to be protected for the rest of their lives. Unlike animal shelters, sanctuaries do not seek to place animals with individuals or groups, instead maintaining each animal until its natural death. However, they can offer rehoming services. In some cases, an establishment may have characteristics of both a sanctuary and a shelter; for instance, some animals may be in residence temporarily until a good home is found and others may be permanent residents. The mission of sanctuaries is generally to be safe havens, where the animals receive the best care that the sanctuaries can provide. Animals are not bought, sold, or traded, nor are they used for animal testing. Additionally, no parts of nor secretions from the animals are commodified, such as eggs, wool, or milk. The resident animals are given the opportunity to behave as natural as possible in a protective environment.

Farm Sanctuary's shelter in upstate New York provides a home to hundreds of rescued goats, sheep, cows, pigs, poultry, rabbits, and other farm animals.

What distinguishes a sanctuary from other institutions is the philosophy that the residents come first. In a sanctuary, every action is scrutinized for any trace of human benefit at the expense of non-human residents. Sanctuaries act on behalf of the animals, and the caregivers work under the notion that all animals in the sanctuary, human and non-human, are of equal importance.

Most sanctuaries are not open to the public in the sense of a zoo; that is, allowing unescorted public access to the facility. A legitimate sanctuary avoids activity that would place the animals in an unduly stressful situation.

One of the most important missions of sanctuaries, beyond caring for the animals, is educating the public. The ultimate goal of many sanctuaries is to change the way that humans think of, and treat, non-human animals.

PROTECTED FOREST

Protected Forests are generally established under individual sub-decrees, specifically for the purpose of protecting biodiversity and conservation. They are home to many endangered or threatened species. The Mondul Kiri Protected Forest, for example, is home to three wild cattle species, cats such as tigers and leopards, sun bears and various primates and birds. There are elephants, bears, and seven species of cat in the Seima Protected Forest.

In addition to their conservation and biodiversity values, protected forests also directly support a number of human communities: Rivers in the Central Cardamoms Protected Forest are the key source of water for over 30,000 people, for example.

Mondul Kiri Protected Forest.

Previously, these forests were managed by the Forest Administration, part of the Ministry of Agriculture, Forestry and Fisheries (MAFF). However, under sub-decree No. 69 on the Transfer of the Protected Forest, Forest Conservation and Production Forest Areas, and Economic Land Concessions between MAFF, and MoE (Ministry of Environment), the protected forests are now under the management of the MoE.

The government also receives technical assistance in its management from bodies such as WWF (Mondul Kiri) the Wildlife Alliance and Conservation International (the Cardamoms) and Wildlife Conservation Society (Seima).

As a result of the subsequent sub-decrees in May 2016, all of the protected forests were then transformed into natural protected areas, except Ta Moa Protected Forest in Takeo province, with a land area of 2,285 hectares. Over a half of these natural protected areas are under the classification of wildlife sanctuary.

Deforestation sometimes occurs in protected forests, although at a much lower rate than surrounding forests. Conservation International estimates a deforestation rate of 2% for the Central Cardamom Mountains Protected Forest over a six-year period.

Both the MoE and MAFF are responsible for cracking down on any illegal logging. While MAFF manages economic land concessions, MoE manages protected forests, forest conservation and production forest areas in addition to the natural protected areas. The transfer of responsibilities should be followed by the allocation of resources to match the new responsibilities.

References

- Wildlife-sanctuary, biology: byjus.com, Retrieved 19 July, 2020
- Dawson, T. P.; Jackson, S. T.; House, J. I.; Prentice, I. C.; Mace, G. M. (2011-04-01). "Beyond Predictions: Biodiversity Conservation in a Changing Climate". Science. 332 (6025): 53–58. Bibcode:2011Sci...332...53D. doi:10.1126/science.1200303. ISSN 0036-8075. PMID 21454781

- Van Dyke, Fred (2008). Conservation Biology: Foundations, Concepts, Applications, 2nd ed. Springer Verlag. p. 478. ISBN 978-1-4020-6890-4
- Tom Regan (2006). Jaulas Vacías. El Desafío de los Derechos de los Animales (in Spanish). Barcelona: Fundación Altarriba. p. 111. ISBN 978-84-611-0672-1
- Protected-forest, topics: opendevelopmentcambodia.net, Retrieved 18 May, 2020

5

Threats to Protected Areas

The most common threats to protected areas include human-wildlife conflict, poaching, skinning, hunting, air pollution and acid rain, etc. This chapter delves into these threats to protected areas for an in-depth understanding of the subject.

Several problems with the world's current protected areas mean that the full diversity of the Earth's habitats and species is not properly protected. These problems include:

- Poor representation of habitats: Many habitats are not well represented in the current network of protected areas. For example, less than 4% of the ocean is protected. Freshwater habitats are also poorly represented.
- Lack of connectivity between protected areas: Some species, especially large animals like cats and bears, need large areas of natural habitat in order to feed and find mates. Few protected areas are large enough to support more than a few individuals of these species, and many are isolated from other areas of natural habitat. To address this, corridors must be put in place between protected areas to allow species to move from one protected habitat to another. The linking of protected areas to form networks or systems is very important for the survival of many species; however, such connectivity remains rare.
- Lack of funds: Putting representative protected area networks in place and managing them effectively requires money. However, few countries, including the richest, have managed to define and establish ways to provide long-term, sustainable financing for individual protected areas, let alone a network. This funding gap is particularly acute in developing countries and for marine protected areas. There is a clear need to find new and sustainable financial resources to supplement funding for existing protected areas and to support the establishment of new protected areas.
- Poor management: The declaration of a protected area is not an end result: a whole series of conditions must be in place for protected areas to be effective. Effective management is essential to ensure that nature is being conserved

within a park's boundaries. Management activities include monitoring the health of habitats, ensuring that the rules of the protected area are respected, and working with local people to balance nature protection with their needs and aspirations.

- Human activities: Closely linked with poor management are threats from widespread and either poorly managed or illegal human activities occurring within protected areas in many parts of the world. These include logging, poaching of protected animals, mining, and encroachment by human settlements and agriculture. Human activities outside of protected areas are also often a threat – such as those leading to pollution, climate change.

Forest elephant killed by poachers for tusks. Poaching is a problem in many protected areas around the world. This forest elephant was killed for its tusks.

HUMAN–WILDLIFE CONFLICT

Grand Canyon National Park hosts millions of visitors every year and is home to a population of Rocky Mountain elk. Interactions between humans and the elk sometimes results in injuries.

Human-wildlife conflict refers to the interaction between wild animals and humans, and the resultant negative impact on people, animals, resources, and habitats. It occurs when growing human populations overlap with established wildlife territory, creating competition for space and resources. Conflict takes many forms including but not limited to: Loss of life or injury to humans and wild animals, depredation of livestock, and degradation of habitat. Human-wildlife conflict is a global issue present in urban and rural landscapes alike.

Previously, conflict mitigation strategies utilized lethal control, translocation, population size regulation, and endangered species preservation. Recent management now uses an interdisciplinary set of approaches to solving conflicts. These include applying scientific research, sociological studies, and the arts to reducing conflicts. As human-wildlife conflict inflicts direct and indirect consequences on people and animals, its mitigation is an important priority for the management of biodiversity and protected areas.

Human–wildlife conflict is defined by the World Wide Fund for Nature (WWF) as "any interaction between humans and wildlife that results in negative impacts on human social, economic or cultural life, on the conservation of wildlife populations, or on the environment. The Creating Co-existence workshop at the 5th Annual World Parks Congress defined human-wildlife conflict in the context of human goals and animal needs as follows: "Human-wildlife conflict occurs when the needs and behavior of wildlife impact negatively on the goals of humans or when the goals of humans negatively impact the needs of wildlife."

A 2007 review by the United States Geological Survey defines human-wildlife conflict in two contexts; firstly, actions by wildlife conflict with human goals i.e. life, livelihood and life-style, and secondly, human activities that threaten the safety and survival of wildlife. However, in both cases outcomes are decided by human responses to the interactions.

The Government of Yukon defines human-wildlife conflict simply, but through the lens of damage to property, i.e. "any interaction between wildlife and humans which causes harm, whether it's to the human, the wild animal, or property." Here, property includes buildings, equipment and camps, livestock and pets, but does not include crops, fields or fences.

Human-wildlife interactions have occurred throughout man's prehistory and recorded history. Among the early forms of human-wildlife conflict is the depredation of the ancestors of prehistoric man by a number of predators of the Miocene such as saber-toothed cats, leopards, and spotted hyenas.

Fossil remains of early hominids show evidence of depredation; the Taung Child, the fossilized skull of a young Australopithecus africanus, is thought to have been killed by an eagle from the distinct marks on its skull and the fossil having been found among egg shells and remains of small animals.

A Plio-Pleistocene horned crocodile, Crocodylus anthropophagus, whose fossil remains have been recorded from Olduvai Gorge, was the largest predator encountered by prehistoric man, as indicated by hominid specimens preserving crocodile bite marks from these sites.

Examples of Human-wildlife Conflict

Asian elephant damages to houses.

As a tropical continent with substantial anthropogenic development, the Africa is a hotspot for biodiversity and therefore, for human-wildlife conflict. Two of the primary examples of conflict in Africa are human-predator (lions, leopards, cheetahs, etc.) and human-elephant conflict. Depredation of livestock by African predators is well documented in Kenya, Namibia, Botswana, and more. African elephants frequently clash with humans, as their long-distance migrations often intersect with farms. The resulting damage to crops, infrastructure, and at times, people, can lead to the retaliatory killing of elephants by locals.

With a rapidly increasing human population and high biodiversity, interactions between people and wild animals are becoming more and more prevalent. Like human-predator in Africa, encounters between tigers, people, and their livestock is a prominent issue on the Asian continent. Attacks on humans and livestock have exacerbated major threats the tiger conservation such as mortality, removal of individuals from the wild, and negative perceptions of the animals from locals. Even non-predator conflicts are common, with crop-raiding by elephants and macaques persisting in both rural and urban environments, respectively. To this date, no instances of human-wildlife conflict have been reported in Antarctica.

Human-wildlife conflict in Europe includes interactions between people and both carnivores and herbivores. A variety of non-predators such as deer, wild boar, rodents, and starlings have been shown to damage crops and forests. Carnivores like raptors and bears create conflict with humans by eating both farmed and wild fish, while others like lynxes and wolves prey upon livestock. Even less apparent cases of human-wildlife conflict can cause substantial losses; 500,000 deer-vehicle collisions in Europe (and 1-1.5 million in North America) led to 30,000 injuries and 200 deaths.

Instances of human-wildlife conflict are widespread in North America. In Wisconsin, United States wolf depredation of livestock is a prominent issue that resulted in the injury or death of 377 domestic animals over a 24-year span. Similar incidents were reported in the Greater Yellowstone ecosystem, with reports of wolves killing pets and livestock. Expanding urban centers have created increasing human-wildlife conflicts, with interactions between human and coyotes and mountain lions documented in cities in Colorado and California, respectively, among others. Big cats are a similar source of conflict in Central Mexico, where reports of livestock depredation are widespread, while interactions between humans and coyotes were observed in Canadian cities as well.

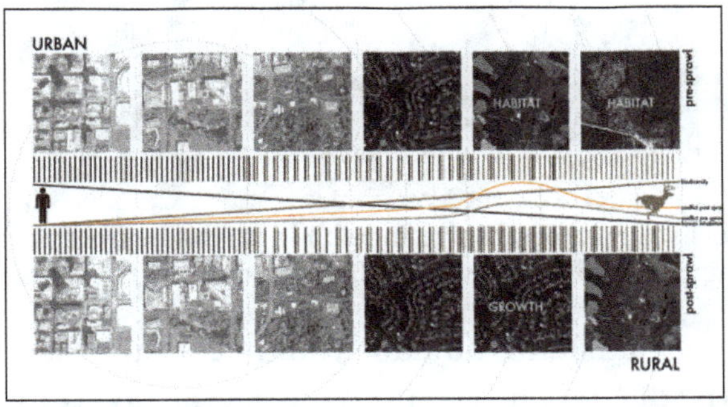

Diagram of Human Wildlife Conflict in Expanding American Cities.

On K'gari-Fraser Island in Australia, attacks by wild dingoes on humans (including the death of a child) created a human-wildlife crisis that required scientific intervention to manage. In New Zealand, distrust and dislike of introducing predatory birds (such as the New Zealand Falcon) to vineyard landscapes led to tensions between people and the surrounding wildlife. In extreme cases large birds have been reported to attack people who approach their nests, with human-magpie conflict in Australia a well-known example. Even conflict in urban environments has been documented, with development increasing the frequency of human-possum interactions in Sydney.

As with most continents, the depredation of livestock by wild animals is a primary source of human-wildlife conflict in South America. The killings of guanacos by predators in Patagonia, Chile – which possess both economic and cultural value in the region – have created tensions between ranchers and wildlife. South America's only species of bear, the Andean Bear, faces population declines due to similar conflict with livestock owners in countries like Ecuador.

Human-wildlife conflict is not limited to terrestrial ecosystems, but is prevalent in the world's oceans as well. As with terrestrial conflict, human-wildlife conflict in aquatic environments is incredibly diverse and extends across the globe. In Hawaii, for example, an increase in monk seals around the islands has created a conflict between locals who believe that seals "belong" and those who do not. Marine predators such as killer whales and fur seals compete with fisheries for food and resources, while others like great white sharks have a history of injuring humans. While many of the causes of human-wildlife conflict are the same between terrestrial and marine ecosystems (depredation, competition, human injury, etc.), ocean environments are less studied and management approaches often differ.

Mitigation Strategies

A traditional livestock corral surrounded by a predator-proof corral in South Gobi desert, Mongolia, to protect livestock from predators like snow leopard and wolf.

Mitigation strategies for managing human-wildlife conflict vary significantly depending on location and type of conflict. The preference is always for passive, non-intrusive prevention measures but often active intervention is required to be carried out in conjunction. Regardless of approach, the most successful solutions are those that include local communities in the planning, implementation, and maintenance. Resolving conflicts, therefore, often requires a regional plan of attack with a response tailored to the specific crisis. Still, there are a variety of management techniques that are frequently employed to mitigate conflicts. Examples include:

- Translocation of problematic animals: Relocating so-called "problem" animals from a site of conflict to a new place is a mitigation technique used in the past, although recent research has shown that this approach can have detrimental impacts on species and is largely ineffective. Translocation can decrease survival rates and lead to extreme dispersal movements for a species, and often "problem" animals will resume conflict behaviors in their new location.

- Erection of fences or other barriers: Building barriers around cattle bomas, creating distinct wildlife corridors, and erecting beehive fences around farms to deter elephants have all demonstrated the ability to be successful and cost-effective strategies for mitigating human-wildlife conflict.

- Improving community education and perception of animals: Various cultures have myriad views and values associated with the natural world, and how wildlife is perceived can play a role in exacerbating or alleviating human-wildlife conflict. In one Masaai community where young men once obtained status by killing lions, conservationists worked with community leaders to shift perceptions and allow those young men to achieve the same social status by protecting lions instead.

- Effective land use planning: Altering land use practices can help mitigate conflict between humans and crop-raiding animals. For example in Mozambique, communities started to grow more chili pepper plants after making the discovery that elephants dislike and avoid plants containing capsaicin. This creative and effective method discourages elephants from trampling community farmers' fields as well as protects the species.

- Compensation: Compensation in some cases, governmental systems have been established to offer monetary compensation for losses sustained due to human-wildlife conflict. These systems hope to deter the need for retaliatory killings of animals, and to financially incentivize the co-existing of humans and wildlife. Compensation strategies have been employed in India, Italy, and South Africa, to name a few. The success of compensation in managing human-wildlife conflict has varied greatly due to under-compensation, a lack of local participation, or a failure by the government to provide timely payments.
- Spatial analyses and mapping conflict hotspots: Mapping interactions and creating spatial models has been successful in mitigating human-carnivore conflict and human-elephant conflict, among others. In Kenya, for example, using grid-based geographical information systems in collaboration with simple statistical analyses allowed conservationists to establish an effective predictor for human-elephant conflict.
- Predator-deterring guard dogs: The use of guard dogs to protect livestock from depredation has been effective in mitigating human-carnivore conflict around the globe. A recent review found that 15.4% of study cases researching human-carnivore conflict used livestock-guarding dogs as a management technique, with animal losses on average 60 times lower than the norm.

Livestock guardian dogs can be an effective and popular way of deterring predators and reducing human-carnivore conflicts.

Hidden Dimensions of Conflict

Human wildlife conflict also has a range of 'hidden' dimensions that are not typically considered when the focus is on visible consequences. These can include health impacts, opportunity costs, and transaction costs. Case studies include work on elephants in northeast India, where human-elephant interactions are correlated with increased imbibing of alcohol by crop guardians with resultant enhanced mortality in interactions, and issues related to gender in northern India. In addition, research has shown that the fear caused by the presence of predators can aggravate human-wildlife conflict more than the actual damage produced by encounters.

POACHING

Poaching has been defined as the illegal hunting or capturing of wild animals, usually associated with land use rights. Poaching was performed by impoverished peasants for subsistence purposes and a supplement for meager diets. It was set against the hunting privileges of nobility and territorial rulers.

Since the 1980s, the term "poaching" has also been used to refer to the illegal harvesting of wild plant species. In agricultural terms, the term 'poaching' is also applied to the loss of soils or grass by the damaging action of feet of livestock which can affect availability of productive land, water pollution through increased runoff and welfare issues for cattle. Stealing livestock as in cattle raiding classifies as theft, not as poaching.

The Poacher.

In 1998, environmental scientists from the University of Massachusetts Amherst proposed the concept of poaching as an environmental crime, defining any activity as illegal that contravenes the laws and regulations established to protect renewable natural resources including the illegal harvest of wildlife with the intention of possessing, transporting, consuming or selling it and using its body parts. They considered poaching as one of the most serious threats to the survival of plant and animal populations. Wildlife biologists and conservationists consider poaching to have a detrimental effect on biodiversity both within and outside protected areas as wildlife populations decline, species are depleted locally, and the functionality of ecosystems is disturbed.

Grave of a poacher in Schliersee, quoting the first stanza of the Jennerwein song. Now and then, poached game is being placed on the grave to commemorate 'Girgl'.

Austria and Germany refer to poaching not as theft, but as intrusion in third party hunting rights. While ancient Germanic law allowed any free man including peasants to hunt, especially on the commons, Roman law restricted hunting to the rulers. In Medieval Europe feudal territory rulers from the king downward tried to enforce exclusive rights of the nobility to hunt and fish on the lands they ruled. Poaching was deemed a serious crime punishable by imprisonment, but the enforcement was comparably weak until the 16th century. Peasants were still allowed to continue small game hunting, but the right of the nobility to hunt was restricted in the 16th century and transferred to land ownership. The low quality of guns made it necessary to approach to the game as close as 30 m (98 ft). For example, poachers in the Salzburg region then were men around 30 years of age, not yet married and usually alone on their illegal trade.

The development of modern hunting rights is closely connected to the comparably modern idea of exclusive private property of land. In the 17th and 18th centuries, the restrictions on hunting and shooting rights on private property were being enforced by gamekeepers and foresters. They denied shared usages of forests, e.g. resin collection and wood pasture and the peasant's right to hunt and fish. However, comparably easy access to rifles increasingly allowed peasants and servants to poach by end of the 18th century. Hunting was used in the 18th century as a theatrical demonstration of the aristocratic rule of the land and had a strong impact on land use patterns as well. Poaching in so far interfered not only with property rights but clashed symbolically with the power of the nobility. During the years between 1830 and 1848 poaching and poaching related deaths increased in Bavaria. The German revolutions of 1848–49 were interpreted as a general allowance for poaching in Bavaria. The reform of hunting law in 1849 reduced legal hunting to rich land owners and the bourgeoisie able to pay the hunting fees and led to disappointment and ongoing praise of poachers among the people. Some of the frontier regions, where smuggling was important, showed especially strong resistance. In 1849, the Bavarian military forces were asked to occupy a number of municipalities on the frontier to Austria. Both, in Wallgau (today a part of Garmisch-Partenkirchen) and in Lackenhäuser in the Bavarian forest each household had to feed and accommodate one soldier for a month as part of a military mission to quell the uproar. The people of Lackenhäuser had several skirmishes about poached deer with Austrian foresters and military, and were known as well armed pertly poachers (kecke Wilderer).

Brass plaque on door at Tremedda farm, warning that poachers shall be shot on first sight.

Poaching, like smuggling, has a long counter-cultural history. The verb *poach* is derived from the Middle English word pocchen literally meaning bagged, enclosed in a bag. Poaching was dispassionately reported for England in "Pleas of the Forest", transgressions of the rigid Anglo-Norman Forest Law. William the Conqueror, who was a great lover of hunting, established and enforced a system of forest law. This operated outside the common law, and served to protect game animals and their forest habitat from hunting by the common people of England and reserved hunting rights for the new French-speaking Anglo-Norman aristocracy. Henceforth hunting of game in royal forests by commoners or in other words poaching, was invariably punishable by death by hanging. In 1087, a poem called "The Rime of King William" contained in the Peterborough Chronicle, expressed English indignation at the severe new laws. Poaching was romanticised in literature from the time of the ballads of Robin Hood, as an aspect of the "greenwood" of Merry England; in one tale, Robin Hood is depicted as offering King Richard the Lion Heart venison from deer illegally hunted in the Sherwood Forest, the King overlooking the fact that this hunting was a capital offence. The widespread acceptance of this common criminal activity is encapsulated in the observation Non est inquirendum, unde venit venison ("It is not to be inquired, whence comes the venison"), made by Guillaume Budé in his Traitte de la vénerie. However, the English nobility and land owners were in the long term extremely successful in enforcing the modern concept of property, expressed e.g. in the enclosures of common land and later in the Highland Clearances, which were both forced displacement of people from traditional land tenancies and erstwhile common land. The 19th century saw the rise of acts of legislation, such as the Night Poaching Act 1828 and Game Act 1831 in the United Kingdom, and various laws elsewhere.

Lady Baltimore, a bald eagle in Alaska survived a poaching attempt in the Juneau Raptor Center mew.

In North America, the blatant defiance of the laws by poachers escalated to armed conflicts with law authorities, including the Oyster Wars of the Chesapeake Bay, and the joint US-British Bering Sea Anti-Poaching Operations of 1891 over the hunting of seals.

Violations of hunting laws and regulations concerning wildlife management, local or international wildlife conservation schemes constitute wildlife crimes that are typically punishable. The following violations and offenses are considered acts of poaching in the USA:

- Hunting, killing or collecting wildlife that is listed as endangered by IUCN and protected by law such as the Endangered Species Act, the Migratory Bird Treaty Act of 1918 and international treaties such as CITES.
- Fishing and hunting without a license.
- Capturing wildlife outside legal hours and outside the hunting season; usually the breeding season is declared as the closed season during which wildlife is protected by law.
- Prohibited use of machine guns, poison, explosives, snare traps, nets and pitfall traps.
- Prohibited use of baiting with food, decoys or recorded calls in order to increase chances for shooting wildlife.

- Hunting from a moving vehicle or aircraft.
- Shining deer with a spotlight at night to impair its natural defenses and thus facilitate an easy kill is considered animal abuse. This hunting method is illegal in California, Virginia, Connecticut, Florida, Michigan and Tennessee.
- Taking wildlife on land that is restricted, owned by or licensed to somebody else.
- The animal or plant has been tagged by a researcher.
- Shooting an animal in a confined area (canned hunting).

Stephen Corry, director of the human-rights group Survival International, has argued that the term "poaching" has at times been used to criminalize the traditional subsistence techniques of indigenous peoples and bar them from hunting on their ancestral lands, when these lands are declared wildlife-only zones. Corry argues that parks such as the Central Kalahari Game Reserve are managed for the benefit of foreign tourists and safari groups, at the expense of the livelihoods of tribal peoples such as the Kalahari bushmen.

Motives

Sociological and criminological research on poaching indicates that in North America people poach for commercial gain, home consumption, trophies, pleasure and thrill in killing wildlife, or because they disagree with certain hunting regulations, claim a traditional right to hunt, or have negative dispositions toward legal authority. In rural areas of the United States, the key motives for poaching are poverty. Interviews conducted with 41 poachers in the Atchafalaya River basin in Louisiana revealed that 37 of them hunt to provide food for themselves and their families; 11 stated that poaching is part of their personal or cultural history; nine earn money from the sale of poached game to support their families; eight feel exhilarated and thrilled by outsmarting game wardens.

In African rural areas, the key motives for poaching are the lack of employment opportunities and a limited potential for agriculture and livestock production. Poor people rely on natural resources for their survival and generate cash income through the sale of bushmeat, which attracts high prices in urban centres. Body parts of wildlife are also in demand for traditional medicine and ceremonies. The existence of an international market for poached wildlife implies that well-organised gangs of professional poachers enter vulnerable areas to hunt, and crime syndicates organise the trafficking of wildlife body parts through a complex interlinking network to markets outside the respective countries of origin. Armed conflict in Africa has been linked to intensified poaching and wildlife declines within protected areas, likely reflecting the disruption of traditional livelihoods, which causes people to seek alternative food sources.

In a study conducted in Tanzania by two scientists named Paul Wilfred and Andrew

Maccoll explored why the people in Tanzania poached certain species and when they are more likely to do so. They decided to interview people from multiple villages who lived near the Ugalla Game Reserve. To make sure the interview and their results were unbiased, they randomly picked several villages and several families from each village to interview.

One of the major cases of poaching is for bushmeat, or meat consumed from non-domesticated species of animals from all sorts of classes such as mammals or birds. Usually, bushmeat is considered a subset of poaching due to the hunting of animals regardless of the laws that conserve certain species of animals. Poachers hunt for bushmeat for both consumption and for profit.

The conclusion of the study found that many families would consume more bushmeat if there weren't protein alternatives such as fish and the further away the families were from the reserve, the less likely they were to illegally hunt the wildlife for bushmeat. Finally, families were more likely to hunt for bushmeat right before harvest season and during heavy rains, as before the harvest season, there is not much agricultural work and heavy rainfall obscures human tracks, making it easier for poachers to get away with their crimes.

Poverty seems to be a large impetus to cause people to poach, something that affects both residents in Africa and Asia. For example, in Thailand, there are anecdotal accounts of the desire for a better life for children, which drive rural poachers to take the risk of poaching even though they dislike exploiting the wildlife.

Another major cause of poaching is due to the cultural high demand of wildlife products, such as ivory, that are seen as symbols of status and wealth in China. According to Joseph Vandegrift, China saw an unusual spike in demand for ivory in the twenty-first century due to the economic boom that allowed more middle-class Chinese to have a higher purchasing power that incentivized them to show off their newfound wealth using ivory, a rare commodity since the Han Dynasty.

In China, there are problems with wildlife conservation, specifically relating to tigers. Several authors collaborated on a piece titled "Public attitude toward tiger farming and tiger conservation in Beijing, China", exploring the option of whether it would be a better policy to raise tigers on a farm or put them in a wildlife conservation habitat to preserve the species. Conducting a survey on 1,058 residents of Beijing, China with 381 being university students and the other 677 being regular citizens, they tried to gauge public opinion about tigers and conservation efforts for them. They were asked questions regarding the value of tigers in relations to ecology, science, education, aestheticism, and culture. However, one reason emerged as to why tigers are still highly demanded in illegal trading: culturally, they are still status symbols of wealth for the upper class, and they are still thought to have mysterious medicinal and healthcare effects.

Effects of Poaching

Memorial to rhinos killed by poachers near.

The detrimental effects of poaching can include:

- Defaunation of forests: Predators, herbivores and fruit-eating vertebrates cannot recover as fast as they are removed from a forest; as their populations decline, the pattern of seed predation and dispersal is altered; tree species with large seeds progressively dominate a forest, while small-seeded plant species become locally extinct.
- Reduction of animal populations in the wild and possible extinction.
- The effective size of protected areas is reduced as poachers use the edges of these areas as open-access resources.
- Wildlife tourism destinations face a negative publicity; those holding a permit for wildlife-based land uses, tourism-based tour and lodging operators lose income; employment opportunities are reduced.
- Emergence of zoonotic diseases caused by transmission of highly variable retrovirus chains:
    - Outbreaks of the Ebola virus in the Congo Basin and in Gabon in the 1990s have been associated with the butchering of apes and consumption of their meat.
    - The outbreak of SARS in Hong Kong is attributed to contact with and consumption of meat from masked palm civets, raccoon dogs, Chinese ferret-badgers and other small carnivores that are available in southern Chinese wildlife markets.
    - Bushmeat hunters in Central Africa infected with the human T-lymphotropic virus were closely exposed to wild primates.
    - Results of research on wild central chimpanzees in Cameroon indicate that they are naturally infected with the simian foamy virus and constitute a reservoir of HIV-1, a precursor of the acquired immunodeficiency syndrome in humans.

Products

A seashell vendor sells seashells to tourists, seashells which have been taken from the sea alive, killing the animal inside.

The body parts of many animals, such as tigers and rhinoceroses, are believed to have certain positive effects on the human body, including increasing virility and curing cancer. These parts are sold in areas where these beliefs are practiced – mostly Asian countries particularly Vietnam and China – on the black market.

Traditional Chinese medicine often incorporates ingredients from all parts of plants, the leaf, stem, flower, root, and also ingredients from animals and minerals. The use of parts of endangered species (such as seahorses, rhinoceros horns, binturong and tiger bones and claws) has created controversy and resulted in a black market of poachers. Deep-seated cultural beliefs in the potency of tiger parts are so prevalent across China and other east Asian countries that laws protecting even critically endangered species such as the Sumatran tiger fail to stop the display and sale of these items in open markets, according to a 2008 report from TRAFFIC. Popular "medicinal" tiger parts from poached animals include tiger genitals, culturally believed to improve virility, and tiger eyes.

A vendor selling illegal items at a Chinese market for use in traditional Chinese medicine. Some of the pieces pictured include parts of animals such as a tiger's paw.

Rhino populations face extinction because of demand in Asia (for traditional medicine and as a luxury item) and in the Middle East (where horns are used for decoration). A sharp surge in demand for rhino horn in Vietnam was attributed to rumors that the horn cured cancer, even though the rumor has no basis in science. Recent prices for a kilo of crushed rhino horn have gone for as much as $60,000, more expensive than a kilo of gold. Vietnam is the only nation which mass-produces bowls made for grinding rhino horn.

Ivory, which is a natural material of several animals, plays a large part in the trade of illegal animal materials and poaching. Ivory is a material used in creating art objects and jewelry where the ivory is carved with designs. China is a consumer of the ivory trade and accounts for a significant amount of ivory sales. In 2012, The New York Times reported on a large upsurge in ivory poaching, with about 70% of all illegal ivory flowing to China.

Fur is also a natural material which is sought after by poachers. A Gamsbart, literally chamois beard, a tuft of hair traditionally worn as a decoration on trachten-hats in the alpine regions of Austria and Bavaria formerly was worn as a hunting (and poaching) trophy. In the past, it was made exclusively from hair from the chamois' lower neck.

Anti-poaching Efforts

TRAFFIC brings to light many of the poaching areas and trafficking routes and helps to clamp down on the smuggling routes the poachers use to get the ivory to areas of high demand, predominantly Asia. As many as 35,000 African elephants are slaughtered yearly to feed the demand for their ivory tusks. This ivory then goes on to be used in jewellery, musical instruments, and other trinkets.

Members of the Rhino Rescue Project have implemented a technique to combat rhino poaching in South Africa by injecting a mixture of indelible dye and a parasiticide into the animals' horns, which enables tracking of the horns and deters consumption of the horn by purchasers. Since rhino horn is made of keratin, advocates say the procedure is painless for the animal. Another strategy being used to counter rhino poachers in Africa is called RhODIS, which is a database that compiles rhino DNA from confiscated horns and other goods that were being illegally traded, as well as DNA recovered from poaching sites. RhODIS cross-references the DNA as it searches for matches; if a match is found, it is used to track down the poachers. Africa's Wildlife Trust seeks to protect African elephant populations from poaching activities in Tanzania. Hunting for ivory was banned in 1989, but poaching of elephants continues in many parts of Africa stricken by economic decline. The International Anti-Poaching Foundation has a structured military-like approach to conservation, employing tactics and technology generally reserved for the battlefield. Founder Damien Mander is an advocate of the use of military equipment and tactics, including Unmanned Aerial Vehicles, for military-style anti-poaching operations. Such military-style approaches have been criticised for failing to resolve the underlying reasons for poaching, but to neither tackle "the role of global trading networks" nor the continued demand for animal products. Instead, they "result in coercive, unjust and counterproductive approaches to wildlife conservation".

Chengeta Wildlife is an organization that works to equip and train wildlife protection teams and lobbies African governments to adopt anti-poaching campaigns. Jim Nyamu's elephant walks are part of attempts in Kenya to reduce ivory poaching.

In 2013, the Tanzanian Minister of Natural Resources and Tourism urged that poachers be shot on sight in an effort to stop the mass killing of elephants. Since December 2016, anti-poaching police units in Namibia are permitted to return fire on poachers if fired upon. The government of Botswana adopted a 'shoot-to-kill' policy against poachers in 2013 as a "legitimate conservation strategy" and "a necessary evil", which has reduced poaching to the point it is thought to be 'virtually non-existent' in the country, and that neighbouring countries like South Africa should also adopt similar measures in order to save wildlife from extinction. In May 2018, the Kenyan government announced that poachers will face the death penalty, as fines and life imprisonment have "not been deterrence enough to curb poaching, hence the proposed stiffer sentence". Human rights organizations oppose the move, but wildlife advocates support it. Save the Rhino, a UK-based wildlife advocacy organization notes that in Kenya, 23 rhinos and 156 elephants were killed by poachers between 2016 and 2017. As of March 2019, the measure is being put on the fast track to implementation by Kenyan lawmakers.

Large quantities of ivory are sometimes destroyed as a statement against poaching (aka "ivory crush"). In 2013 the Philippines were the first country to destroy their national seized ivory stock. In 2014 China followed suit and crushed six tons of ivory as a symbolic statement against poaching.

There are two main solutions according to Frederick Chen that would attack the supply side of this poaching problem to reduce its effects: enforcing and enacting more policies and laws for conservation and by encouraging local communities to protect the wildlife around them by giving them more land rights.

Nonetheless, Frederick Chen wrote about two types of effects stemming from demand-side economics: The bandwagon and snob effect. The former deals with people desiring a product due to many other people buying it, while the latter is similar but with one distinct difference: people will clamour to buy something if it denotes wealth that only a few elites could possibly afford. Therefore, the snob effect would offset some of the gains made by anti-poaching laws, regulations, or practices: if a portion of the supply is cut off, the rarity and price of the object would increase, and only a select few would have the desire and purchasing power for it. While approaches to dilute mitigate poaching from a supply-side may not be the best option as people can become more willing to purchase rarer items, especially in countries gaining more wealth and therefore higher demand for illicit goods—Frederick Chen still advocates that we should also focus on exploring ways to reduce the demand for these goods to better stop the problem of poaching. Indeed, there is some evidence that interventions to reduce consumer demand may be more effective for combatting poaching than continually increased policing to catch poachers. However, almost no groups deploying interventions that attempt to reduce consumer demand evaluate the impact of their actions.

Another solution to alleviate poaching proposed in Tigers of the World was about how to implement a multi-lateral strategy that targets different parties to conserve wild tiger populations in general. This multi-lateral approach include working with different agencies to fight and prevent poaching since organized crime syndicates benefit from tiger poaching and trafficking; therefore, there is a need to raise social awareness and implement more protection and investigative techniques. For example, conservation groups raised more awareness amongst park rangers and the local communities to understand the impact of tiger poaching—they achieved this through targeted advertising that would impact the main audience. Targeting advertising using more violent imagery to show the disparity between tigers in nature and as a commodity made a great impact on the general population to combat poaching and indifference towards this problem. The use of spokespeople such as Jackie Chan and other famous Asian actors and models who advocated against poaching also helped the conservation movement for tigers too.

In July 2019, rhino horns encased in plaster were seized in Vietnam that were being trafficked from the United Arab Emirates. Despite the ban on trade since the 1970s, poaching level of rhino horns has risen over the last decade, leading the rhino population into crisis.

Poaching has many causes in both Africa and China. The issue of poaching is not a simple one to solve as traditional methods to counter poaching have not taken into the account the poverty levels that drive some poachers and the lucrative profits made by organized crime syndicates who deal in illegal wildlife trafficking. Conservationists hope the new emerging multi-lateral approach, which would include the public, conservation groups, and the police, will be successful for the future of these animals.

Some game wardens have made use of robotic decoy animals placed in high visibility areas to draw out poachers for arrest after the decoys are shot. Decoys with robotics to mimic natural movements are also in use by law enforcement.

SKINNING

Skinning is the act of skin removal. The process is done with animals, mainly as a means to prepare the muscle tissues beneath for consumption or for use of the fur or tanning of the skin. The skin may also be used as a trophy, sold on the fur market, or, in the case of a declared pest, used as proof of kill to obtain a bounty from a government health, agricultural, or game agency.

A Kalanga man skinning a goat at the annual Domboshaba cultural festival.

Two common methods of skinning are open skinning and case skinning. Typically, large animals are open skinned and smaller animals are case skinned. Skinning, when it is performed on live humans as a form of capital punishment or murder, is referred to as flaying.

Skinning Methods

Case skinning is a method where the skin is peeled from the animal like a sock. This method is usually used if the animal is going to be stretched out or put in dry storage. Many smaller animals are case skinned, leaving the skin mostly undamaged in the shape of a tube.

Although the methods of case skinning individual animals vary slightly, the general steps remain the same. To case skin an animal, it is hung upside down by its feet. A cut is made in one foot, and continued up the leg, around the anus and down the other leg. From there the skin is pulled down the animal as though removing a sweater.

Open skinning is a method where the skin is removed from the animal like a jacket. This method is generally used if the skin is going to be tanned immediately or frozen for storage. A skin removed by the open method can be used for wall hangings or rugs. Larger animals are often skinned using the open method. To open skin an animal, the body is placed on a flat surface. A cut is made from the anus to the lower lip, and up the legs of the animal. The skin is then opened and removed from the animal. The final step is to scrape the excess fat and flesh from the inside of the skin with a blunt stone or bone tool.

Dorsal skinning is very similar to open skinning, however instead of making a cut up the stomach of the animal, the cut is made along the spine. This method of skinning is very popular among taxidermists, as the backbone is easier to access and cleaner than the stomach and between the legs. A dorsal incision is made by laying the animal on its abdomen and making a single cut from the base of the tail to the shoulder region. The animal's skin is easier to remove if the animal has been freshly killed.

Cape skinning is the process of removing the shoulder, neck and head skin for the purpose of displaying the animal as a trophy.

Animal Skins and Native Americans

Native Americans used skins for many purposes other than decoration, clothing and blankets. Animal skin was a staple in the Native Americans' daily lives. It was used to make tents, to build boats, to make bags, to create musical instruments such as drums, and to make quivers.

Since Native Americans were practiced in the means of acquiring and manipulating animal skin, fur trading developed from contact between them and Europeans in the 16th century. Animal skin was a valuable currency which the Native Americans had in excess and would trade for things such as iron-based tools and tobacco which were common in the more developed European areas. Beaver hats became very popular towards the end of the 16th century, and skinning beavers was necessary to acquire their wool. In this time, the beaver skin drastically rose in demand and in value. However, the high number of beavers being harvested for their pelts led to a depletion of beavers, and the industry had to slow down.

HUNTING

Hunting is the practice of seeking, pursuing and capturing or killing wild animals. Hunting wildlife or feral animals is most commonly done by humans for food, recreation, to remove predators that can be dangerous to humans or domestic animals, to remove pests that destroy crops or kill livestock, or for trade. Lawful hunting is distinguished from poaching, which is the illegal killing, trapping or capture of the hunted species. The species that are hunted are referred to as game or prey and are usually mammals and birds. In economic terms, hunting is considered to be part of the primary production alongside forestry, agriculture and fishing.

Hunter on the stand during a driven hunt.

Hunting arose in Homo erectus or earlier, on the order of millions of years ago. Hunting is deeply embedded in human culture. Hunting an animal for its meat can also be seen as a more natural way to obtain animal protein since regulated hunting does not cause the same environmental issues as raising domestic animals for meat, especially on factory farms.

Hunting can also be a means of pest control. Hunting advocates state that hunting can be a necessary component of modern wildlife management, for example, to help maintain a population of healthy animals within an environment's ecological carrying capacity when natural checks such as predators are absent or very rare. However, the usefulness of hunting as a control measure has been questioned, and excessive hunting has also heavily contributed to the endangerment, extirpation and extinction of many animals.

Bushmen hunter.

The pursuit, capture and release, or capture for food of fish is called fishing, which is not commonly categorised as a form of hunting. It is also not considered hunting to pursue animals without intent to kill them, as in wildlife photography, birdwatching, or scientific research activities which involve tranquilizing or tagging of animals or birds. The practice of foraging or gathering materials from plants and mushrooms is also considered separate from hunting.

Skillful tracking and acquisition of an elusive target has caused the word hunt to be used in the vernacular as a metaphor, as in treasure hunting, "bargain hunting", and even "hunting down" corruption and waste. Some animal rights activists argue that hunting is cruel, unnecessary, and unethical.

Hunting has a long history. It pre-dates the emergence of Homo sapiens (anatomically modern humans) and may even predate genus Homo.

The oldest undisputed evidence for hunting dates to the Early Pleistocene, consistent with the emergence and early dispersal of Homo erectus, about 1.7 million years ago (Acheulean). While it is undisputed that Homo erectus were hunters, the importance of this for the emergence of Homo erectus from its australopithecine ancestors, including the production of stone tools and eventually the control of fire, is emphasised in the so-called "hunting hypothesis" and de-emphasised in scenarios that stress omnivory and social interaction.

Professional deer stalker with red stag.

There is no direct evidence for hunting predating Homo erectus, in either Homo habilis or in Australopithecus. The early hominid ancestors of humans were probably frugivores or omnivores, with a partially carnivore diet from scavenging rather than hunting. Evidence for australopithecine meat consumption was presented in the 1990s. It has nevertheless often been assumed that at least occasional hunting behavior may have been present well before the emergence of Homo. This can be argued on the basis of comparison with chimpanzees, the closest extant relatives of humans, who also engage in hunting, indicating that the behavioral trait may have been present in the Chimpanzee–human last common ancestor as early as 5 million years ago. The common chimpanzee (Pan troglodytes) regularly engages in troop predation behaviour where bands of beta males are led by an alpha male. Bonobos (Pan paniscus) have also been observed to occasionally engage in group hunting, although more rarely than Pan troglodytes, mainly subsisting on a frugivorous diet. Indirect evidence for Oldowan era hunting, by early Homo or late Australopithecus, has been presented in a 2009 study based on an Oldowan site in southwestern Kenya.

Louis Binford criticised the idea that early hominids and early humans were hunters. On the basis of the analysis of the skeletal remains of the consumed animals, he concluded that hominids and early humans were mostly scavengers, not hunters, Blumenschine proposed the idea of confrontational scavenging, which involves challenging and scaring off other predators after they have made a kill, which he suggests could have been the leading method of obtaining protein-rich meat by early humans.

Stone spearheads dated as early as 500,000 years ago were found in South Africa. Wood does not preserve well, however, and Craig Stanford, a primatologist and professor of anthropology at the University of Southern California, has suggested that the discovery of spear use by chimpanzees probably means that early humans used wooden spears as well, perhaps, five million years ago. The earliest dated find of surviving wooden hunting spears dates to the very end of the Lower Paleolithic, just before 300,000 years ago. The Schöningen spears, found in 1976 in Germany, are associated with Homo heidelbergensis.

The hunting hypothesis sees the emergence of behavioral modernity in the Middle Paleolithic as directly related to hunting, including mating behaviour, the establishment of language, culture, and religion, mythology and animal sacrifice.

Evidence exists that hunting may have been one of the multiple environmental factors leading to the Holocene extinction of megafauna and their replacement by smaller herbivores. North American megafauna extinction was coincidental with the Younger Dryas impact event, possibly making hunting a less critical factor in prehistoric species loss than had been previously thought. However, in other locations such as Australia, humans are thought to have played a very significant role in the extinction of the Australian megafauna that was widespread prior to human occupation.

Saharan rock art with prehistoric archers.

Inuit walrus hunters.

Hunting was a crucial component of hunter-gatherer societies before the domestication of livestock and the dawn of agriculture, beginning about 11,000 years ago in some parts of the world. In addition to the spear, hunting weapons developed during the Upper Paleolithic include the atlatl a spear-thrower and the bow. By the Mesolithic, hunting strategies had diversified with the development of these more far-reaching weapons and the domestication of the dog about 15,000 years ago. Evidence puts the earliest known mammoth hunting in Asia with spears to approximately 16,200 years ago.

Sharp flint piece from Bjerlev Hede in central Jutland.

Many species of animals have been hunted throughout history. It has been suggested that in North America and Eurasia, caribou and wild reindeer "may well be the species of single greatest importance in the entire anthropological literature on hunting", although the varying importance of different species depended on the geographic location.

Mesolithic hunter-gathering lifestyles remained prevalent in some parts of the Americas, Sub-Saharan Africa, and Siberia, as well as all of Australia, until the European Age of Discovery. They still persist in some tribal societies, albeit in rapid decline. Peoples that preserved Paleolithic hunting-gathering until the recent past include some indigenous peoples of the Amazonas (Aché), some Central and Southern African (San people), some peoples of New Guinea (Fayu), the Mlabri of Thailand and Laos, the Vedda people of Sri Lanka, and a handful of uncontacted peoples. In Africa, one of the last remaining hunter-gatherer tribes are the Hadza of Tanzania.

Ancient Greek black-figure pottery depicting the return of a hunter and his dog; made in Athens.

Even as animal domestication became relatively widespread and after the development of agriculture, hunting was usually a significant contributor to the human food supply. The supplementary meat and materials from hunting included protein, bone for implements, sinew for cordage, fur, feathers, rawhide and leather used in clothing.

Hunting is still vital in marginal climates, especially those unsuited for pastoral uses or agriculture. For example, Inuit people in the Arctic trap and hunt animals for clothing and use the skins of sea mammals to make kayaks, clothing, and footwear.

On ancient reliefs, especially from Mesopotamia, kings are often depicted as hunters of big game such as lions and are often portrayed hunting from a war chariot. The cultural and psychological importance of hunting in ancient societies is represented by deities such as the horned god Cernunnos and lunar goddesses of classical antiquity, the Greek Artemis or Roman Diana. Taboos are often related to hunting, and mythological association of prey species with a divinity could be reflected in hunting restrictions such as a reserve surrounding a temple. Euripides' tale of Artemis and Actaeon, for example, may be seen as a caution against disrespect of prey or impudent boasting.

Artemis with a Hind, a Roman copy of an Ancient Greek sculpture.

With the domestication of the dog, birds of prey, and the ferret, various forms of animal-aided hunting developed, including venery (scent hound hunting, such as fox hunting), coursing (sight hound hunting), falconry, and ferreting. While these are all associated with medieval hunting, over time, various dog breeds were selected for very precise tasks during the hunt, reflected in such names as pointer and setter.

An example of a Goguryeo tomb mural of hunting, middle of the first millennium.

Pastoral and Agricultural Societies

Even as agriculture and animal husbandry became more prevalent, hunting often remained as a part of human culture where the environment and social conditions allowed. Hunter-gatherer societies persisted, even when increasingly confined to marginal areas. And within agricultural systems, hunting served to kill animals that prey upon domestic and wild animals or to attempt to extirpate animals seen by humans as competition for resources such as water or forage.

When hunting moved from a subsistence activity to a social one, two trends emerged:

- The development of the role of the specialist hunter, with special training and equipment.
- The co-option of hunting as a "sport" for those of an upper social class.

The meaning of the word game in Middle English evolved to include an animal which is hunted. As game became more of a luxury than a necessity, the stylised pursuit of it also became a luxury. Dangerous hunting, such as for lions or wild boars, often done on horseback or from a chariot, had a function similar to tournaments and manly sports. Hunting ranked as an honourable, somewhat competitive pastime to help the aristocracy practice skills of war in times of peace.

In most parts of medieval Europe, the upper class obtained the sole rights to hunt in certain areas of a feudal territory. Game in these areas was used as a source of food and furs, often provided via professional huntsmen, but it was also expected to provide a form of recreation for the aristocracy. The importance of this proprietary view of game can be seen in the Robin Hood legends, in which one of the primary charges against the outlaws is that they "hunt the King's deer". In contrast, settlers in Anglophone colonies gloried democratically in hunting for all. In Medieval Europe, hunting was considered by Johannes Scotus Eriugena to be part of the set of seven mechanical arts.

Use of Dog

Hunting Companions, Dutch 19th-century painting featuring two dogs, a shotgun and a game bag.

Although various other animals have been used to aid the hunter, such as ferrets, the dog has assumed many very important uses to the hunter. The domestication of the dog has led to a symbiotic relationship in which the dog's independence from humans is deferred. Though dogs can survive independently of humans, and in many cases do, as with feral dogs, where hunger is not a primary factor, the species tends to defer to human control in exchange for habitation, food and support.

Dogs today are used to find, chase, retrieve, and sometimes to kill the game. Hunting dogs allow humans to pursue and kill prey that would otherwise be very difficult or dangerous to hunt. Different breeds of dogs are used for different types of hunting. Waterfowl are commonly hunted using retrieving dogs such as the Labrador Retriever, the Golden Retriever, the Chesapeake Bay Retriever, the Brittany Spaniel, and other similar breeds. Game birds are flushed out using flushing spaniels such as the English Springer Spaniel, the various Cocker Spaniels and similar breeds.

The hunting of wild mammals in England and Wales with dogs was banned under the Hunting Act 2004. The wild mammals include fox, hare, deer and mink. Hunting with dogs is permissible, however, where it has been carried out in accordance with one of the exceptions in the Act.

Wildlife Management

Control fence to assess the impact of browsing by ungulates – note the lack of natural forest regeneration outside the fencing.

Hunting is claimed to give resource managers an important tool in managing populations that might exceed the carrying capacity of their habitat and threaten the well-being of other species, or, in some instances, damage human health or safety.

In some cases, hunting actually can increase the population of predators such as coyotes by removing territorial bounds that would otherwise be established, resulting in excess neighbouring migrations into an area, thus artificially increasing the population. Hunting advocates assert that hunting reduces intraspecific competition for food and shelter, reducing mortality among the remaining animals. Some environmentalists assert that reintroducing predators would achieve the same end with greater efficiency and less negative effect, such as introducing significant amounts of free lead into the environment and food chain.

In the United States, wildlife managers are frequently part of hunting regulatory and licensing bodies, where they help to set rules on the number, manner and conditions in which game may be hunted.

Management agencies sometimes rely on hunting to control specific animal populations, as has been the case with deer in North America. These hunts may sometimes be carried out by professional shooters, although others may include amateur hunters. Many US city and local governments hire professional and amateur hunters each year to reduce populations of animals such as deer that are becoming hazardous in a restricted area, such as neighbourhood parks and metropolitan open spaces.

A large part of managing populations involves managing the number and, sometimes, the size or age of animals harvested so as to ensure the sustainability of the population. Tools that are frequently used to control harvest are bag limits and season closures, although gear restrictions such as archery-only seasons are becoming increasingly popular in an effort to reduce hunter success rates in countries that rely on bag limits per hunter instead of per area.

Methods

Africans hunting the lion.

American bison being chased off a cliff as seen and painted by Alfred Jacob Miller.

Historical, subsistence, and sport hunting techniques can differ radically, with modern hunting regulations often addressing issues of where, when, and how hunts are conducted. Techniques may vary depending on government regulations, a hunter's personal ethics, local custom, hunting equipment, and the animal being hunted. Often a hunter will use a combination of more than one technique. Laws may forbid sport hunters from using some methods used primarily in poaching and wildlife management.

- Baiting is the use of decoys, lures, scent, or food.
- Battue involves scaring animals (by beating sticks) into a killing zone or ambush.
- Beagling is the use of beagles in hunting rabbits, and sometimes in hunting foxes.
- Beating uses human beaters to flush out game from an area or drive it into position.

- Stand hunting or blind hunting is waiting for animals from a concealed or elevated position, for example from tree stands, hunting blinds or other types of shooting stands.
- Calling is the use of animal noises to attract or drive animals.
- Camouflage is the use of visual or odour concealment to blend with the environment.
- Dogs may be used to course or to help flush, herd, drive, track, point at, pursue, or retrieve prey.
- Driving is the herding of animals in a particular direction, usually toward another hunter in the group.
- Flushing is the practice of scaring animals from concealed areas.
- Ghillie suit is a type of gear a person can wear to blend with environment.
- Glassing is the use of optics, such as binoculars, to locate animals more easily.
- Glue is an indiscriminate passive form to kill birds.
- Internet hunting is a method of hunting over the Internet using webcams and remotely controlled guns.
- Netting involves using nets, including active netting with the use of cannon nets and rocket nets.
- Persistence hunting is the use of running and tracking to pursue the prey to exhaustion.
- Posting is done by sitting or standing in a particular place with the intentions of intercepting your game of choice along their travel corridor.
- Scouting for game is typically done prior to a hunt and will ensure the desired species are in a chosen area. Looking for animal sign such as tracks, scat, etc. and utilizing “trail cameras” are commonly used tactics while scouting.
- Shooting is the use of a ranged weapon such as a gun, bow, crossbow, or slingshot.
- Solunar theory says that animals move according to the location of the moon in comparison to their bodies and is said to have been used long before this by hunters to know the best times to hunt their desired game.
- Spotlighting or shining is the use of artificial light to find or blind animals before killing.

- Stalking or still hunting is the practice of walking quietly in search of animals or in pursuit of an individual animal.
- Tracking is the practice of reading physical evidence in pursuing animals.
- Trapping is the use of devices such as snares, pits, and deadfalls to capture or kill an animal.

Economics

Chatelherault, built by William Adam as the Duke of Hamilton's hunting lodge.

A variety of industries benefit from hunting and support hunting on economic grounds. In Tanzania, it is estimated that a safari hunter spends fifty to one hundred times that of the average ecotourist. While the average photo tourist may seek luxury accommodation, the average safari hunter generally stays in tented camps. Safari hunters are also more likely to use remote areas, uninviting to the typical ecotourist. Advocates argue that these hunters allow for anti-poaching activities and revenue for local communities.

In the United Kingdom, the game hunting of birds as an industry is said to be extremely important to the rural economy. The Cobham Report of 1997 suggested it to be worth around £700 million, and hunting and shooting lobby groups claimed it to be worth over a billion pounds less than ten years later.

Hunting also has a significant financial impact in the United States, with many companies specialising in hunting equipment or speciality tourism. Many different technologies have been created to assist hunters, even including iPhone applications. Today's hunters come from a broad range of economic, social, and cultural backgrounds. In 2001, over thirteen million hunters averaged eighteen days hunting, and spent over $20.5 billion on their sport. In the US, proceeds from hunting licenses contribute to state game management programs, including preservation of wildlife habitat.

Environmental Problems

Lead bullets that miss their target or remain in an unretrieved carcass could become a toxicant in the environment but lead in ammunition because of its metallic form has a lower solubility and higher resistance to corrosion than other forms of lead making it hardly available to biological systems. Waterfowl or other birds may ingest the lead and poison themselves with the neurotoxicant, but studies have demonstrated that effects of lead in ammunition are negligible on animal population size and growth. Since 1991, US federal law forbids lead shot in waterfowl hunts, and 30 states have some type of restriction.

Right: 40 S and W round with hollow-point bullet, Left: Expanded bullet of the same calibre with exposed lead core.

In December 2014, a federal appeals court denied a lawsuit by environmental groups that the EPA must use the Toxic Substances Control Act to regulate lead in shells and cartridges. The groups sought EPA to regulate "spent lead", yet the court found EPA could not regulate spent lead without also regulating cartridges and shells.

Conservation

Punishment of a Hunter.

Hunters have been driving forces throughout history in the movement to ensure the preservation of wildlife habitats and wildlife for further hunting. However, excessive hunting and poachers have also contributed heavily to the endangerment, extirpation and extinction of many animals, such as the quagga, the great auk, Steller's sea cow, the thylacine, the bluebuck, the Arabian oryx, the Caspian and Javan tigers, the markhor, the Sumatran rhinoceros, the bison, the North American cougar, the Altai argali sheep, the Asian elephant and many more, primarily for commercial sale or sport. All these animals have been hunted to endangerment or extinction. Poaching currently threatens bird and mammalian populations around the world.

Species

Arabian Oryx

The Arabian oryx, a species of large antelope, once inhabited much of the desert areas of the Middle East. However, the species' striking appearance made it (along with the closely related scimitar-horned oryx and addax) a popular quarry for sport hunters, especially foreign executives of oil companies working in the region. The use of automobiles and high-powered rifles destroyed their only advantage: Speed, and they became extinct in the wild exclusively due to sport hunting in 1972. The scimitar-horned oryx followed suit, while the addax became critically endangered. However, the Arabian oryx has now made a comeback and been upgraded from "extinct in the wild" to "vulnerable" due to conservation efforts like captive breeding.

Markhor

The markhor is an endangered species of wild goat which inhabits the mountains of Central Asia and Pakistan. The colonization of these regions by Britain gave British sport hunters access to the species, and they were hunted heavily, almost to the point of extinction. Only their willingness to breed in captivity and the inhospitability of their mountainous habitat prevented this. Despite these factors, the markhor is still endangered.

American Bison

The American bison is a large bovid which inhabited much of western North America prior to the 1800s, living on the prairies in large herds. However, the vast herds of bison attracted market hunters, who killed dozens of bison for their hides only, leaving the rest to rot. Thousands of these hunters quickly eliminated the bison herds, bringing the population from several million in the early 1800s to a few hundred by the 1880s. Conservation efforts have allowed the population to increase, but the bison remains near-threatened due to lack of habitat.

White Rhino

The Journal of International Wildlife Law and Policy cites that the legalization of white rhinoceros hunting in South Africa motivated private landowners to reintroduce the species onto their lands. As a result, the country saw an increase in white rhinos from fewer than one hundred individuals to more than 11,000, even while a limited number were killed as trophies.

However, the illegal hunting of rhinoceros for their horns is highly damaging to the population and is currently growing globally, with 1004 being killed in South Africa alone according to the most recent estimate.

Other Species

According to Richard Conniff, Namibia is home to 1,750 of the roughly 5,000 black rhinos surviving in the wild because it allows trophy hunting of various species. Namibia's mountain zebra population has increased to 27,000 from 1,000 in 1982. Elephants, which "are gunned down elsewhere for their ivory", have gone to 20,000 from 15,000 in 1995. Lions, which were on the brink of extinction "from Senegal to Kenya", are increasing in Namibia.

In contrast, Botswana has recently been forced to ban trophy hunting following a precipitous wildlife decline. The numbers of antelope plummeted across Botswana, with a resultant decline in predator numbers, while elephant numbers remained stable and hippopotamus numbers rose. According to the government of Botswana, trophy hunting is at least partly to blame for this, but many other factors, such as poaching, drought and habitat loss are also to blame. Uganda recently did the same, arguing that "the share of benefits of sport hunting were lopsided and unlikely to deter poaching or improve capacity to manage the wildlife reserves."

Studies

Cage trap (live trap) for cheetahs on a farm.

A study issued by the Wildlife Society concluded that hunting and trapping are cost effective tools that reduce wildlife damage by reducing a population below the capacity of the environment to carry it and changing the behaviors of animals to stop them from causing damage. The study furthermore states that the cessation of hunting could cause wildlife to be severely harmed, rural property values to fall, and the incentive of landowners to maintain natural habitats to diminish.

Opposition to Hunting

Animal rights activists state that killing animals for sport is unethical, cruel, and unnecessary. They note the suffering and cruelty inflicted on animals hunted for sport: "Many animals endure prolonged, painful deaths when they are injured but not killed by hunters. Hunting disrupts migration and hibernation patterns and destroys families." Animal rights activists also comment that hunting is not needed in order to maintain an ecological balance, and that "nature takes care of its own". They say that hunting can be combated on public lands by "spread deer repellent or human hair (from barber shops) near hunting areas". Animal rights activists also argue that hunting is speciesist:

> "Whether hunters try to justify their killing by citing human deaths caused by wild animals, by making conservationist claims, by claiming that it's acceptable to hunt as long as the animals' bodies are eaten, or simply because of the pleasure it brings them, the fact remains that hunting is morally unacceptable if we consider the interests of nonhuman animals. Hunted animals endure fear and pain, and then are deprived of their lives. Understanding the injustices of speciesism and the interests of nonhuman animals makes it clear that human pleasure cannot justify nonhuman animals' pain."

CRUELTY TO ANIMALS

Domitian, one of the 17 macaque monkeys experimented on during the animal experiment in Silver Spring.

Cruelty to animals, also called animal abuse, animal neglect or animal cruelty, is the infliction by omission (animal neglect) or by commission by humans of suffering or harm upon any non-human. More narrowly, it can be the causing of harm or suffering for specific achievement, such as killing animals for food, their fur or even their tusks; opinions differ about the extent of cruelty associated with a given method of slaughter. Cruelty to animals sometimes encompasses inflicting harm or suffering as an end in itself, defined as zoosadism. With approximately 65 billion animals killed annually for food, farm animals are the most numerous animals subjected to cruelty.

Divergent approaches to laws concerning animal cruelty occur in different jurisdictions throughout the world. For example, some laws govern methods of killing animals for food, clothing, or other products, and other laws concern the keeping of animals for entertainment, education, research, or pets. There are a number of conceptual approaches to the issue of cruelty to animals.

Some think that the animal welfare position holds that there is nothing inherently wrong with using animals for human purposes, such as food, clothing, entertainment, fun and research, but that it should be done in a way that minimizes unnecessary pain and suffering, sometimes referred to as "humane" treatment. Others have argued that the definition of 'unnecessary' varies widely and could include virtually all current use of animals.

Utilitarian advocates argue from the position of costs and benefits and vary in their conclusions as to the allowable treatment of animals. Some utilitarians argue for a weaker approach which is closer to the animal welfare position, whereas others argue for a position that is similar to animal rights. Animal rights theorists criticize these positions, arguing that the words "unnecessary" and "humane" are subject to widely differing interpretations, and that animals have basic rights. They say that most animal use itself is unnecessary and a cause of suffering, so the only way to ensure protection for animals is to end their status as property and to ensure that they are never used as a substance or as a non-living thing.

Man beating a chained pitbull terrier with a strap.
The strap is visible in the foreground.

Throughout history, humans believed in a God-given right to treat nonhuman animals with cruelty; however, some individuals, like Leonardo da Vinci for example, who once purchased caged birds in order to set them free, were concerned. His notebooks also record his anger with the fact that humans used their dominance to raise animals for slaughter. According to contemporary philosopher Nigel Warburton, for most of human history the dominant view has been that animals are there for humans to do with as they see fit.

René Descartes believed that non-humans are automata complex machines with no soul, mind, or reason. In Cartesian dualism, consciousness was unique to human among all other animals and linked to physical matter by divine grace. However, close analysis shows that many human features such as complex sign usage, tool use, and self-consciousness can be found in some animals.

Charles Darwin, by presenting the theory of evolution, revolutionized the way that humans viewed their relationship with other species. Darwin believed that not only did human beings have a direct kinship with other animals, but the latter had social, mental and moral lives too. Later, in The Descent of Man, he wrote: "There is no fundamental difference between man and the higher mammals in their mental faculties."

Modern philosophers and intellectuals, such as Peter Singer and Tom Regan, have argued that animals' ability to feel pain as humans do makes their well-being worthy of equal consideration. There are many precursors of this train of thought. Jeremy Bentham, the founder of utilitarianism, famously wrote in his An Introduction to the Principles of Morals and Legislation:

> "The question is not, can they reason nor can they talk? but, can they suffer?"

These arguments have prompted some to suggest that animals' well-being should enter a social welfare function directly, not just indirectly via its effect only on human well-being. Many countries have now formally recognized animal sentience and animal suffering, and have passed anti-cruelty legislation in response.

Forms

A starved and bruised horse eating at a veterinary clinic after rescue.

Animal cruelty can be broken down into two main categories: active and passive. Passive cruelty is typified by cases of neglect, in which the cruelty is a lack of action rather than the action itself. Oftentimes passive animal cruelty is accidental, born of ignorance. In many cases of neglect in which an investigator believes that the cruelty occurred out of ignorance, the investigator may attempt to educate the pet owner, then revisit the situation. In more severe cases, exigent circumstances may require that the animal be removed for veterinary care.

Industrial Animal Farming

Farm animals are generally produced in large, industrial facilities that house thousands of animals at high densities; these are sometimes called factory farms. The industrial nature of these facilities means that many routine procedures or animal husbandry practices impinge on the welfare of the animals and could be considered as cruelty, with Henry Stephen Salt claiming in 1899 that "it is impossible to transport and slaughter vast numbers of large and highly-sensitive animals in a really humane manner". It has been suggested the number of animals hunted, kept as companions, used in laboratories, reared for the fur industry, raced, and used in zoos and circuses, is insignificant compared to farm animals, and therefore the "animal welfare issue" is numerically reducible to the "farm animal welfare issue". Similarly, it has been suggested by campaign groups that chickens, cows, pigs, and other farm animals are among the most numerous animals subjected to cruelty. For example, because male chickens do not lay eggs, newly hatched males are culled using macerators or grinders. Worldwide meat overconsumption is another factor that contributes to the miserable situation of farm animals. Many undercover investigators have exposed the animal cruelty taking place inside the factory farming industry and there is evidence to show that consumers provided with accurate information about the process of meat productions and the abuse that accompanies it has led to changes in their attitudes.

A chicken egg production facility.

The American Veterinary Medical Association accepts maceration subject to certain conditions, but recommends alternative methods of culling as more humane. Egg-laying hens are then transferred to "battery cages" where they are kept in high densities. Matheny and Leahy attribute osteoporosis in hens to this caging method. Broiler chickens suffer similar situations, in which they are fed steroids to grow at a super-fast speed, so fast that their bones, heart and lungs often cannot keep up. Broiler chickens under six weeks old suffer painful crippling due to fast growth rates, whilst one in a hundred of these very young birds dies of heart failure.

To reduce aggression in overcrowded conditions, shortly after birth piglets are castrated, their tails are amputated, and their teeth clipped. Calves are sometimes raised in veal crates, which are small stalls that immobilize calves during their growth, reducing costs and preventing muscle development, making the resulting meat a pale color, preferred by consumers.

Egg laying hens in a crowded cage.

Animal cruelty such as soring, which is illegal, sometimes occurs on farms and ranches, as does lawful but cruel treatment such as livestock branding. Since Ag-gag laws prohibit video or photographic documentation of farm activities, these practices have been documented by secret photography taken by whistleblowers or undercover operatives from such organizations as Mercy for Animals and the Humane Society of the United States posing as employees. Agricultural organizations such as the American Farm Bureau Federation have successfully advocated for laws that tightly restrict secret photography or concealing information from farm employers.

Alleged Link to Human Violence and Psychological Disorders

There are studies providing evidence of a link between animal cruelty and violence towards humans. A 2009 study found that slaughterhouse employment increases total arrest rates, arrests for violent crimes, arrests for rape, and arrests for other sex offenses in comparison with other industries.

A history of torturing pets and small animals, a behavior known as zoosadism, is considered one of the signs of certain psychopathologies, including antisocial personality disorder, also known as psychopathic personality disorder. According to The New York Times, "the FBI has found that a history of cruelty to animals is one of the traits that regularly appears in its computer records of serial rapists and murderers, and the standard diagnostic and treatment manual for psychiatric and emotional disorders lists cruelty to animals a diagnostic criterion for conduct disorders." "A survey of psychiatric patients who had repeatedly tortured dogs and cats found all of them had high levels of aggression toward people as well, including one patient who had murdered a young boy." Robert K. Ressler, an agent with the Federal Bureau of Investigation's behavioral sciences unit, studied serial killers and noted, "Murderers like this very often start out by killing and torturing animals as kids."

Acts of intentional animal cruelty or non-accidental injury may be indicators of serious psychological problems. According to the American Humane Association, 13% of intentional animal abuse cases involve domestic violence. As many as 71% of pet-owning women seeking shelter at safe houses have reported that their partner had threatened and hurt or killed one or more of their pets; 32% of these women reported that one or more of their children had also hurt or killed pets. Battered women report that they are prevented from leaving their abusers because they fear what will happen to the animals in their absence. Animal abuse is sometimes used as a form of intimidation in domestic disputes.

Cruelty to animals is one of the three components of the Macdonald triad, behavior considered to be one of the signs of violent antisocial behavior in children and adolescents. According to the studies used to form this model, cruelty to animals is a common (but not universal) behavior in children and adolescents who grow up to become serial killers and other violent criminals. It has also been found that children who are cruel to animals have often witnessed or been victims of abuse themselves. In two separate studies cited by the Humane Society of the United States, roughly one-third of families suffering from domestic abuse indicated that at least one child had hurt or killed a pet.

Cultural Rituals

Many times, when Asiatic elephants are captured in Thailand, handlers use a technique known as the training crush, in which "handlers use sleep-deprivation, hunger, and thirst to 'break' the elephants' spirit and make them submissive to their owners"; moreover, handlers drive nails into the elephants' ears and feet.

The practice of cruelty to animals for divination purposes is found in ancient cultures, and some modern religions such as Santeria continue to do animal sacrifices for healing and other rituals. Taghairm was performed by ancient Scots to summon devils.

Circuses

The use of animals in the circus has been controversial since animal welfare groups have documented instances of animal cruelty during the training of performing animals. Numerous instances of animal abuse in circuses have been documented such as confining enclosures, lack of regular veterinary care, abusive training methods and lack of oversight by regulating bodies. Animal trainers have argued that some criticism is not based on fact, including beliefs that shouting makes the animals believe the trainer is going to hurt them, that caging is cruel and common, and the harm caused by the use of whips, chains or training implements.

Some circuses now present animal-free acts. Bolivia has enacted what animal rights activists called the world's first ban on all animals in circuses.

Bullfighting

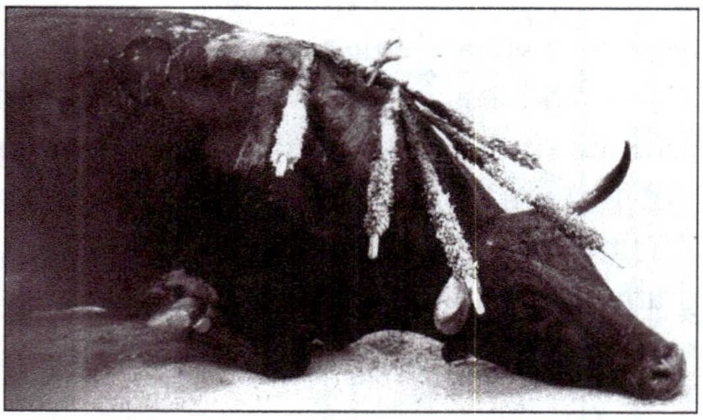

A bull dying in a bullfight.

Bullfighting is criticized by animal rights or animal welfare activists, referring to it as a cruel or barbaric blood sport in which the bull suffers severe stress and a slow, torturous death. A number of activist groups undertake anti-bullfighting actions in Spain and other countries. In Spanish, opposition to bullfighting is referred to as antitaurismo.

The Bulletpoint Bullfight warns that bullfighting is "not for the squeamish", advising spectators to "be prepared for blood". It details prolonged and profuse bleeding caused by horse-mounted lancers, the charging by the bull of a blindfolded, armored horse who is "sometimes doped up, and unaware of the proximity of the bull", the placing of barbed darts by banderilleros, followed by the matador's fatal sword thrust. It stresses that these procedures are a normal part of bullfighting and that death is rarely instantaneous. It further warns those attending bullfights to "be prepared to witness various failed attempts at killing the animal before it lies down."

Toro Embolado

The Toro Jubilo.

The "Toro Jubilo" or Toro embolado in Soria, Medinaceli, Spain, is a festival associated with animal cruelty. During this festival, balls of pitch are attached to a bull's horns and set on fire. The bull is then released into the streets and can do nothing but run around in pain, often smashing into walls in an attempt to douse the fire. These fiery balls can burn for hours, and they burn the bull's horns, body, and eyes – all while spectators cheer and run around the victim. The animal rights group PACMA has described the fiesta as "a clear example of animal mistreatment", and PETA calls it "a sadistic festival".

A horse with a gas mask during World War I.

Rattlesnake Round-ups

Rattlesnake round-ups, also known as rattlesnake rodeos, are annual events common in the rural Midwest and Southern United States, where the primary attractions are captured wild rattlesnakes which are sold, displayed, killed for food or animal products (such as snakeskin) or released back into the wild. The largest rattlesnake round-up in the United States is held in Sweetwater, Texas. Held every year since 1958, the event currently attracts approximately 30,000 visitors per year and in 2006 each annual round-up was said to result in the capture of 1% of the state's rattlesnake population. Rattlesnake round-ups became a concern by animal welfare groups and conservationists due to claims of animal cruelty and excessive threat of future endangerment. In response, some round-ups impose catch-size restrictions or releasing captured snakes back into the wild.

Warfare

Military animals are creatures that have been employed by humankind for use in warfare. They are a specific application of working animals. Examples include horses, dogs and dolphins. Only recently has the involvement of animals in war been questioned, and practices such as using animals for fighting, as living bombs (as in the use of exploding donkeys) or for military testing purposes (such as during the Bikini atomic experiments) may now be criticised for being cruel.

Princess Anne, the Princess Royal, the patron of the British Animals in War Memorial, stated that animals adapt to what humans want them to do, but that they will not do things that they do not want to, even with training. Animal participation in human conflict was commemorated in the United Kingdom in 2004 with the erection of the Animals in War Memorial in Hyde Park, London.

In 2008 a video of US Marine David Motari throwing a puppy over a cliff during the Iraq conflict was popularised as an internet phenomenon and attracted widespread criticism of the soldier's actions for being an act of cruelty.

Unnecessary Scientific Experiments or Demonstrations

Under all three of the conceptual approaches to animal cruelty discussed above, performing unnecessary experiments or demonstrations upon animals that cause them substantial pain or distress may be viewed as cruelty. Due to changes in ethical standards, this type of cruelty tends to be less common today than it used to be in the past. For example, schoolroom demonstrations of oxygen depletion routinely suffocated birds by placing them under a glass cover, and animals were suffocated in the Cave of Dogs to demonstrate the density and toxicity of carbon dioxide to curious travellers on the Grand Tour.

No Pet Policies and Abandonment

Many apartment complexes and rental homes institute no pet policies. No pet policies are a leading cause of animal abandonment, which is considered a crime in many jurisdictions. In many cases, abandoned pets have to be euthanized due to the strain they put on animal shelters and rescue groups. Abandoned animals often become feral or contribute to feral populations. In particular, feral dogs can pose a serious threat to pets, children, and livestock. Feral cats are known to hunt endangered birds and have interbred with wild cat populations.

In Ontario, Canada, no pet policies are outlawed under the Ontario Landlord and Tenant Act and are considered invalid even when a tenant signs a lease that includes a no pets clause. Similar legislation has also been considered in Manitoba.

HABITAT ENCROACHMENT

Encroachment is the spread of humans, transportation systems, utilities, buildings, and other development into natural areas. Encroachment leads to loss of habitat or changes in the natural ecology for plants, animals, fish, and other living organisms that live in the area. Whether it is due to development, hunting, mining, agriculture, or migrating populations, habitat encroachment increasingly exposes vulnerable humans to altered ecosystems and to new diseases.

Animals need access to food, water, and shelter to live and even plants need the natural conditions in soil, weather, and rainfall to which they have adapted. But with habitat encroachment, the ecosystems that support life are altered or eliminated. For example, draining a lake of all its water will kill the fish and other organisms living in the water and affect the health or migration patterns of birds or mammals that rely on the water for survival. Changes in the ecosystem can lead to overpopulation of some plants and animals that thrive in the new conditions and declining numbers of those that relied on the original environment. Most habitat encroachment can be attributed to increasing human populations and urban sprawl.

Biodiversity keeps natural systems in balance, so that different forms of life can survive and support one another. This means eventually humans are affected by their actions of encroaching on habitats. For example, if people continue to use all the fresh water, they will have less available to drink, and they will see declines in populations of fish for consumption. In general, human activities such as encroachment that lead to declining or extinction of species hurt crops and other food sources. Indigenous people are particularly threatened by these changes because they rely on local forests, grasslands, or other ecosystems for their survival.

One type of encroachment is habitat fragmentation. When roads, structures, or agricultural and logging operations divide a natural area such as grassland, it is considered fragmented. Division of the original area into smaller and sometimes isolated parcels are called habitat islands. Results can be loss of tree cover or disruption of animal migration routes. Fragmentation can also divide animal populations into smaller groups, making them more vulnerable to predators or disease. Animals might have trouble locating food and water. Sometimes, encroachment introduces new species of plants or animals, which then become invasive.

The ecosystems on Earth, which have evolved over millions of years, support the diverse and complicated forms of life on the planet. Since the earliest reports of human life, the diverse ecosystems have provided food, energy, water, and even medicine for people. In more recent history, especially in the twentieth century, the natural sources for plants, animals, and people have changed and declined on Earth. In some areas, such as Europe, hundreds of years of grazing livestock and harvesting meadows have led to encroachment on local forests and vegetation. These practices have caused an imbalance in the types of plants growing in any given area and an imbalance in the animal life the land supports. In the 1990s, about 94 million acres of global forest area were lost because people converted the forest to agricultural use or urban settlement. This loss amounted to nearly 2.4% of total forests.

Demographics

Causes and Symptoms

Logging and unsustainable agricultural practices, such as deforestation for monoculture as in the palm oil industry in Indonesia and subsistence farming, are among the leading causes of habitat loss. Others are mining, climate change, and pollution. Overexploitation, which is hunting and fishing in an unsustainable way, also adds to habitat degradation. Development in suburban areas also causes encroachment. Developers clear land of trees and other vegetation to build homes or commercial properties. Building along shorelines often includes erecting walls or jetties to protect the shoreline, but they also affect wave energy and sediment layers nearby.

Deforestation: Borneo tropical rainforest is destroyed for oil palm plantations and human development.

Even when habitats are not entirely lost, they can be altered so much by human encroachment that they fail to support natural life. The encroachment can lead to lack of food or drinking water, causing diseases of the stomach, along with undernutrition, malnutrition, or dehydration. Humans also affect habitats by accidentally introducing organisms not native to the area. For example, bringing plants native to other parts of the world into a new community can lead to new pests that affect crops, or the introduction of invasive plants that choke out native plants essential for the local habitat. Declining populations of plants or animals in the area typically are the first symptoms of habitat encroachment. Logging or other removal of vegetation can lead to increased erosion, affecting nearby water flow and quality. Agricultural practices might cause runoff of pesticides and livestock waste, which can pollute freshwater supplies for people and fish and introduce diseases.

Common diseases and disorders:

- A fungus has been introduced in wetlands around the world that is causing disease in amphibians and deleting their populations.
- Introduction of giant snails from East Africa to replace more expensive escargot in Brazil spread rapidly and depleted native plants and food. The snails also introduced life-threatening form of meningitis among locals.
- A probable link between declining tropical forest habitats from encroachment and the spread of the *Ebola* virus exists. Bats are considered the reservoir for the virus, which is then transmitted to humans and other primates. The largest outbreak of Ebola virus disease in history, a 2014 outbreak originating in West Africa, is thought to have started when a child living in a small village in the

Gueckedou district of Guinea, a heavily forested region disturbed by intermittent mining and logging, played near a tree where carrier bats were living.

- A link between encroachment on forest areas and Lyme disease when human development caused natural predators of mice to leave, adding to the mouse population and ticks that carry the disease.

Treatment

Prevention of encroachment on habitats and sustainable practices in hunting, farming, logging, and development can stop many diseases and other health issues from occurring. Given that only about 1% of animal viruses are known to people, outbreaks and death in humans and animals can occur before treatments or preventing measures are developed. For example, when the Ebola fever outbreak occurred in 2014, there was no treatment for the disease in people, and as of 2018 no treatment was available. Doctors can only help relieve symptoms and results of the fever.

Public Health Role and Response

Many of the health issues from habitat encroachment can be prevented or lessened with adequate laws and enforcement of laws. For example, control of hunting and fishing practices, and especially poaching, can slow encroachment on and decline of native species. The U.S. Endangered Species Act, introduced in 1973 in the United States, was designed to prevent declining populations of individual species of animals but was not very successful. A more comprehensive approach might be to introduce laws protecting ecosystems and habitats that support several species and people. International treaties have been signed by some countries to help maintain ecosystems and protect native species.

Prevention

People around the world, especially those heading industries that use natural resources, are being encouraged to join in ensuring sustainable, long-term approaches to dealing with habitat encroachment, and eventually prevent disease outbreaks and other health effects of encroachment. Communities and individuals can support seed banks, which preserve seeds from endangered plant species, and wildlife farms that offer habitats for animals.

Individuals can help prevent unnecessary damage to habitats by learning more about local natural habitats, by reducing pollution, and by planting native species on their properties to provide habitats where encroachment has begun. It also helps to consider when building on one's property or supporting development in a community the possible impact of these actions on habitats. Everyone can be aware of and discourage introduction of invasive plant or animal species. Communities can work with public health officials to minimize the chance that diseases and other health effects of habitat encroachment become outbreaks and serious public health crises.

AIR POLLUTION AND ACID RAIN

Animals are exposed to air pollutants via three pathways: 1) Inhalation of gases or small particles; 2) Ingestion of particles suspended in food or water; or 3) Absorption of gases through the skin. In general, only soft-bodied invertebrates (e.g. earthworms), or animals with thin, moist skin (e.g. amphibians) are affected by the absorption of pollutants. An individual's response to a pollutant varies greatly and depends on the type of pollutant involved, the duration and time of exposure, and the amount taken up by the animal. The individual's age, sex, health, and reproductive condition also play a role in its response. There is a great deal of variability between animal classes, species, and even genotypes, in terms of the level of tolerance to a particular pollutant. The pollutants of concern have been divided into three broad categories: Gases, such as ozone and hydrogen sulphide; non-acidic particulates and toxins, like metals, fluoride compounds, and organic and synthetic chemicals; and acidifying agents, specifically nitrates and sulphates.

Gaseous Pollutants

Volatile organic compounds and nitrogen oxides, emitted from industrial processes, undergo chemical transformation in the atmosphere in the presence of sunlight to form ozone. Ozone, sulphur dioxide, and nitrogen dioxide primarily affect the respiratory system, and it is likely that birds are even more susceptible to gaseous pollutant injury than mammals due to their higher respiratory rates.

Non-acidic Particulates and Toxins

There are a number of air pollutants that are categorised as particulates. Heavy metals (e.g. lead, arsenic, and cadmium) are emitted by smelters; fluoride is emitted in both gaseous and particulate form from aluminium reduction plants and coal-fired power plants; and dioxins, furans, and mercury are emitted by resource recovery facilities. Metals may affect the circulatory, respiratory, gastrointestinal, and central nervous systems of animals. Often organs such as the kidney, liver, and brain are targeted. Entire populations can be affected as metal contamination can cause changes in birth, growth, and death rates.

Fluoride poisoning, or fluorosis, causes gross malformations of bones and teeth. Plants take up gaseous fluoride and store it in their tissues, and fluoride in particulate form is deposited on leaf surfaces and stays there until washed off. Herbivores are best known for exhibiting symptoms of fluoride poisoning. However, earthworms and other soil invertebrates also accumulate fluoride, which is, in turn, passed on to the animals that eat them.

Organic and synthetic chemicals, such as dioxins and organochlorines, affect wildlife. Dioxins bioaccumulate, or build up in the body by concentrating in body fat, and they are resistant to biological breakdown. A study of earthworms showed they accumulated dioxin up to five times the concentration found in the soil, although this dose was not lethal to the worms. Nevertheless, this non-lethal accumulation could have strong ecological implications, since earthworms are a major source of food for a number of bird and small mammal species, many of which have exhibited carcinogenic, reproductive, and immunotoxic effects after exposure to low levels of dioxins.

Nitrates and Sulphates

Sulphur dioxide and nitrogen oxides emitted as a result of fossil fuel combustion undergo chemical transformation in the atmosphere, and occur as sulphate, nitrate, and hydrogen ions when dissolved in precipitation known as "acid rain". Well-buffered soils can adsorb sulphate and neutralise acidity, resulting in soil water and streamwater composition being maintained in a range acceptable to organisms. The adsorption capacity of even well-buffered soils is limited, however, and long-term deposition of acidic compounds depletes the supply of base cations in the soils that buffer these inputs. The build-up of sulphates and nitrates in soils can result in delayed acidification of surface waters once saturation is reached in sensitive watersheds.

The effects of decreasing pH on aquatic invertebrates and fish have been summarised in a National Acid Precipitation Assessment Program (NAPAP) report. Insect taxa differ greatly in their response to acidity, with some species affected at pH levels near 6.0. In the early stages of acidification, acid-sensitive species are replaced by acid-tolerant ones. However, as pH levels continue to drop, more species are lost.

Many studies have demonstrated that surface water acidification can lead to a decline in, and loss of, fish populations. Below pH 4.5 no fish are likely to survive. Fish loss is occurring in many countries, including Scandinavia, Scotland, Wales and North America. A decrease in pH is often associated with an increase in metal availability, being particularly true for aluminium and mercury. Decreased pH and elevated aluminium have been shown to increase fish mortality, decrease fish growth, decrease egg production and embryo survival, and result in physiological impairment of adult fish. In general, embryos, fry, and juveniles are less acid-tolerant than adult fish. Aluminium can precipitate onto fish gills, inhibiting diffusion and resulting in respiratory stress.

Acid deposition is a possible cause of declines in amphibian populations. The larval stages of aquatic amphibian species are most affected by acidic water. Many frog species use temporary ponds, but these tend to be small and shallow, and are easily affected by precipitation chemistry because their only sources of water are rainfall and snowmelt. Frogs that use large, permanent bodies of water for breeding generally lay their eggs in the summer, so they do not experience the acid pulses from snowmelt. However, the eggs and larvae of these species are even more sensitive to subtle changes in pH levels than those of species that breed in the temporary ponds. As is true with fish, the toxic effect of decreased pH levels on amphibians is complicated when concentrations of metals, such as aluminium, in the water increase, but as a general rule, embryos of sensitive amphibian species are killed by water with a pH of 4.5 or lower, while embryos of tolerant species can survive down to a pH of 3.7.

Indirect Effects

In addition to affecting individual animals or populations directly, air pollutants also affect wildlife indirectly by causing changes in the ecosystem. Vegetation affords cover for protection from predators and weather, provides breeding and nesting habitat, and also serves as a food source. Therefore, any change in vegetation could indirectly affect animal populations. Many studies have found that invertebrates show a preference for, or are better able to establish themselves in, air pollution-injured vegetation.

Fluoride and heavy metals can accumulate in the soil to levels that are toxic to soil invertebrates. Species sensitive to metals are replaced by ones that are more metal-tolerant. For example, soft-bodied species such as earthworms and nematodes seem to be more readily affected by elevated metal concentrations. Invertebrates play an important role in forest floor litter decomposition. As forest floor litter builds up, mineral release is delayed, and the availability of nutrients to plants is reduced. Herbivores are ultimately affected when the quantity or quality of their food supply decreases.

References

- Protected-area-problems, biodiversity-protected-areas: wwf.panda.org, Retrieved 17 March, 2020
- "Poaching". Random House Webster's Unabridged Dictionary (2nd ed.). New York: Random House. 2002. ISBN 978-0-375-42599-8
- Habitat-Encroachment, environmental-health: reference.jrank.org, Retrieved 16 April, 2020
- Madden, Francine (2004). "Creating Coexistence between Humans and Wildlife: Global Perspectives on Local Efforts to Address Human–Wildlife Conflict" (PDF). Human Dimensions of Wildlife. Taylor & Francis Inc. 9 (4): 247–257. doi:10.1080/10871200490505675. ISSN 1087-1209. Retrieved 17 July, 2013
- Peterson, M. Nils (2019), Fath, Brian D. (ed.), "Hunting", Encyclopedia of Ecology (2 ed.), Elsevier, 3, pp. 438–440, doi:10.1016/b978-0-12-409548-9.11168-6, ISBN 978-0-444-64130-4, retrieved 20 January, 2020
- Over-exploitation-is-the-biggest-threat-to-wildlife-20160818: traveller24.com, Retrieved 22 June, 2020
- Warchol, G. (2004). "The Transnational Illegal Wildlife Trade". Criminal Justice Studies. 17 (1): 57–73. doi:10.1080/08884310420001679334

Permissions

All chapters in this book are published with permission under the Creative Commons Attribution Share Alike License or equivalent. Every chapter published in this book has been scrutinized by our experts. Their significance has been extensively debated. The topics covered herein carry significant information for a comprehensive understanding. They may even be implemented as practical applications or may be referred to as a beginning point for further studies.

We would like to thank the editorial team for lending their expertise to make the book truly unique. They have played a crucial role in the development of this book. Without their invaluable contributions this book wouldn't have been possible. They have made vital efforts to compile up to date information on the varied aspects of this subject to make this book a valuable addition to the collection of many professionals and students.

This book was conceptualized with the vision of imparting up-to-date and integrated information in this field. To ensure the same, a matchless editorial board was set up. Every individual on the board went through rigorous rounds of assessment to prove their worth. After which they invested a large part of their time researching and compiling the most relevant data for our readers.

The editorial board has been involved in producing this book since its inception. They have spent rigorous hours researching and exploring the diverse topics which have resulted in the successful publishing of this book. They have passed on their knowledge of decades through this book. To expedite this challenging task, the publisher supported the team at every step. A small team of assistant editors was also appointed to further simplify the editing procedure and attain best results for the readers.

Apart from the editorial board, the designing team has also invested a significant amount of their time in understanding the subject and creating the most relevant covers. They scrutinized every image to scout for the most suitable representation of the subject and create an appropriate cover for the book.

The publishing team has been an ardent support to the editorial, designing and production team. Their endless efforts to recruit the best for this project, has resulted in the accomplishment of this book. They are a veteran in the field of academics and their pool of knowledge is as vast as their experience in printing. Their expertise and guidance has proved useful at every step. Their uncompromising quality standards have made this book an exceptional effort. Their encouragement from time to time has been an inspiration for everyone.

The publisher and the editorial board hope that this book will prove to be a valuable piece of knowledge for students, practitioners and scholars across the globe.

Index

www.ingramcontent.com/pod-product-compliance
Lightning Source LLC
LaVergne TN
LVHW080628210926
848485LV00003B/229